Principles of
Health and Safety
at Work

by

ALLAN ST.JOHN HOLT BA FIOSH RSP MAmerSSE
HUW ANDREWS BSc(Aston) FIOSH RSP MAmerSSE
(HASCOM NETWORK LIMITED)

FOREWORD

Developed originally as an in-house workbook by Courtaulds plc for their managers who had line responsibility for health and safety at work, this book quickly established itself as the essential primer for everyone studying for the National Certificate of the National Examination Board in Occupational Safety and Health (NEBOSH).

It has now been updated and extended to cover emergent EC-driven legislation and the resulting practical implications of compliance therewith, thus ensuring that it remains not only the essential NEBOSH Certificate primer but also an invaluable reference and self tuition manual for all with responsibilities for safety and health at work. The book is highly recommended.

J.M.Totterdell M.B.E.
President
The Institution of Occupational Safety and Health (IOSH)

INTRODUCTION TO THE SECOND EDITION

This Second Edition of "Principles" has been written to coincide with the introduction of the 'six-pack' set of Regulations on health and safety at work resulting from EC Directives. It is, we believe, accurate at 16th December 1992. No firm information was available at that time about the details of the 'companion' Regulations on Product and Machinery Safety with the same date of introduction, or the Fire provisions of the Workplace Directive. Readers are strongly advised to acquaint themselves with these as they become available, since there are many cross-links between them and the Regulations discussed in this book.

Sections covering pre-1974 legislation remain in the text because of their continuing applicability in certain circumstances during the various transition periods. Where necessary, they have been revised to delete Regulations and Sections unequivocally revoked.

Two other new Sections have been added, on Ergonomics and Risk Assessment, in response to requests and extension of legal requirements. Others have been extensively rewritten. Our objective, to present complex material in a plain and practical style, remains. We hope that "Principles 2" will be an even more useful desk reference than its predecessor; the mistakes are ours.

Allan St.John Holt
Huw Andrews

First Edition: June 1991
Supplement: March 1992
Second Edition: January 1993
Third impression

ISBN 0 901 357 17 0

Copyright 1993 IOSH Publishing Ltd.
Printed in England by The Cavendish Press Ltd

Contents

Module A: Safety Technology

1.	Safe machinery design and guarding	2
2.	Mechanical handling	6
3.	Manual handling	9
4.	Access equipment	13
5.	Transport safety	19
6.	Classification of dangerous substances	22
7.	Chemical safety	24
8.	Electricity and electrical equipment	27
9.	Fire	31
10.	Construction safety	36
11.	Demolition	40
12.	Selected references	43

Module B: Occupational Health and Hygiene

1.	Introduction to occupational health and hygiene	48
2.	Body response	50
3.	Routes of entry	52
4.	Occupational exposure limits	53
5.	Environmental monitoring	55
6.	Environmental engineering controls	58
7.	Noise	60
8.	Personal protective equipment	63
9.	Radiation	67
10.	Ergonomics	71
11.	Selected references	73

Module C: Safety Management Techniques

1.	Accident prevention	76
2.	Techniques of safety management	79
3.	Risk assessment	83
4.	Safety policy, organisation and the legal framework	86
5.	Safe systems of work	88
6.	Health and safety training	90
7.	Maintenance	93
8.	Management and control of contractors	95
9.	Accident investigation, recording and analysis	98
10.	Information sources	101
11.	Communicating safety	103
12.	Techniques of inspection	105
13.	Selected references	110

Contents

Module D: Law

1.	Introduction to the Law Module	114
2.	The English legal system	115
3.	Common and statute law	119
4.	The Health and Safety at Work etc Act 1974	122
5.	The Management of Health and Safety at Work Regulations 1992	126
6.	The Workplace (Health, Safety and Welfare) Regulations 1992	128
7.	The Provision and Use of Work Equipment Regulations 1992	132
8.	The Manual Handling Operations Regulations 1992	135
9.	The Health and Safety (Display Screen Equipment) Regulations 1992	137
10.	The Personal Protective Equipment at Work Regulations 1992	140
11.	The Factories Act 1961	142
12.	The Offices, Shops and Railway Premises Act 1963	145
13.	The Fire Precautions Act 1971	148
14.	The Control of Substances Hazardous to Health Regulations 1988	150
15.	The Electricity at Work Regulations 1989	153
16.	The Noise at Work Regulations 1989	156
17.	The Abrasive Wheels Regulations 1970	158
18.	The Health and Safety (First-Aid) Regulations 1981	161
19.	The Reporting of Injuries, Diseases and Dangerous Occurrences Regulations 1985	163
20.	Introduction to the Construction Regulations 1961 and 1966	165
21.	The Construction (General Provisions) Regulations 1961	167
22.	The Construction (Lifting Operations) Regulations 1961	171
23.	The Construction (Working Places) Regulations 1966	174
24.	The Construction (Health and Welfare) Regulations 1966	177
25.	The Construction (Head Protection) Regulations 1989	179
26.	The Highly Flammable Liquids and Liquefied Petroleum Gases Regulations 1972	181
27.	The Safety Representatives and Safety Committees Regulations 1977	184
28.	The Safety Signs Regulations 1980	187
29.	The Ionising Radiations Regulations 1985	189
30.	The Control of Pesticides Regulations 1986	191
31.	The Health and Safety Information for Employees Regulations 1989	193
32.	The social security system and benefits	194

Self-Assessment Questions and Answers

Module A	196
Module B	199
Module C	202
Module D	208

Module A:
Safety Technology

INTRODUCTION

Dangers from machinery can arise in two physical ways. The first is possibly the easier to recognise – MACHINERY HAZARDS, including traps, impact, contact, entanglement or ejection of parts of and by the machine, or failure of components. The second way is by NON-MACHINERY HAZARDS, which include electrical failure, exposure to chemical sources, pressure, temperature, noise, vibration and radiation. Danger can also arise from the "software element" – computer control and human intervention by the person carrying out the task at the machine. This Section covers machinery hazards.

It is important to distinguish between continuing dangers associated with the normal working of the machine, such as not having guards fitted where necessary to protect the operator, and those arising from the failure of components or safety mechanisms, such as breakdown of the guarding mechanism.

RISK ASSESSMENT

The amount of risk depends upon the circumstances. These include the type of machine and what it is used for, the need for approach to it, ease of access, and the quality of supervision of the operator's behaviour. Risk also depends upon the knowledge, skills and attitude of the person(s) present in those circumstances, and an individual's awareness of the danger and the skill needed to avoid it. The ability to identify those at risk and to identify when the risk occurs is important, and applies to management as well as to operators.

METHODS OF PREVENTING MACHINERY ACCIDENTS

Machinery safety can be achieved by:

a) eliminating the cause of the danger ("intrinsic safety", see below)
b) reducing or eliminating the need for people to approach the dangerous part(s) of the machine
c) making access to the dangerous parts difficult (or providing safety devices so that access does not lead to injury)
d) the provision of protective clothing or equipment.

(These are listed in order of preference, and may be used in combination).

The safety of operators and those nearby can be achieved by:

a) training to improve peoples' ability to recognise danger
b) redesigning to make dangers more obvious (or use warning signs)
c) training to improve skills in avoiding injury
d) improving motivation to take the necessary avoiding action.

SAFETY BY DESIGN

INTRINSIC SAFETY: A process by which the designer eliminates dangers at the design stage with consideration for the elimination of dangerous parts, making parts inaccessible, reducing the need to handle work pieces in the danger areas, provision of automatic feed devices, and enclosure of the moving parts in the machine.

CONTROL SELECTION: The design and provision of controls which are:

- in the correct position
- of the correct type
- remove the risk of accidental start-up
- have a directional link (control movement matched to machinery movement)
- distinguished by direction of movement
- possessed of distinguishing features (size, colour, feel etc.)

FAILURE TO SAFETY: Designers should ensure that machines fail to safety and not to danger.

Examples of this are clutches scotched to prevent operation, provision of arrestor devices to prevent unexpected strokes and movement, fitting of catches and fall-back devices, and fail-safe electrical limit switches.

MAINTENANCE AND ISOLATION PROCEDURES: Designers should consider the safety of operatives during cleaning and maintenance operations. Routine adjustments, lubrication etc. should be carried out without the removal of safeguards or dismantling of machinery components. Where frequent access is required, interlock guards can be used. Access equipment, where essential, should be provided, as long as access to dangerous parts is prevented. Self-lubrication of parts should be considered if access is difficult. Positive lock-off devices

Safe Machinery Design and Guarding

should be provided to prevent unintentional restarting of machinery. Permit-to-work systems (See Module C Section 5) will be required in some circumstances.

SAFETY BY POSITION: Parts of machines which are out of reach and kept out of reach are called safe by position. However, designers must consider the likelihood of dangerous parts normally out of reach becoming accessible in some circumstances. An example of when this could happen is during painting of a factory machinery area from ladders.

MACHINE LAYOUT: The way in which machines are arranged in the workplace can reduce accidents significantly. A safe layout will take account of:

- Spacing – to facilitate access for operation, supervision, maintenance, adjustment and cleaning
- Lighting – both general lighting to the workplace (natural or artificial but avoiding glare) and localised for specific operations at machines
- Cables and pipes – should be placed to allow safe access and to avoid tripping, with sufficient headroom
- Safe access for maintenance

MACHINE GUARDING

Many serious accidents at work involve the use of machinery. In circumstances where intrinsic safety (see above) has not been achieved, machinery guarding will be the final option. Whether a particular guard is effective or not will depend on its design, and the way in which it relates to the operating procedures and what the machine is used for.

People can be injured directly by machinery in FIVE different ways. Some machines can injure in more than one way. These are:

TRAPS: The body or limb(s) become trapped between closing or passing motions of the machine. In some cases the trap occurs when the limb(s) are drawn into a closing motion, for example in-running nips.

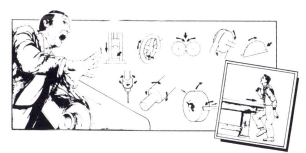

IMPACT: Injuries can result from being struck by moving parts of the machinery.

CONTACT: Injuries can result from contact of the operator with sharp or abrasive surfaces. Alternatively, contact with hot or electrically live components will cause injury.

ENTANGLEMENT: Injuries resulting from the entanglement of hair, rings, items of clothing in moving (particularly rotating) parts of machinery.

EJECTION: Injuries can result from elements of the workpiece or components of machinery being thrown out during the operation of the machine, for example sparks, swarf, chips, molten metal splashes and broken components.

METHODS OF ELIMINATING DANGER THROUGH MACHINERY GUARDING

In circumstances where intrinsic safety through the design is not achieved, machinery guarding will be required to eliminate the dangers. The selection of the material from which the guards will be constructed is determined by FOUR main considerations. These are:

- strength, stiffness and durability
- effects on machine reliability (for example, a solid guard may cause the machine to overheat)
- visibility (there may be operational and safety reasons for needing a clear view of the danger area)
- the control of other hazards (for example, limiting the amount of noise output by choice of special materials).

ERGONOMICS AND THE DESIGN OF MACHINERY GUARDS

Guards should be designed with people in mind. The study of the relationship between differing body shapes and the requirements for reaching and vision in the safe operation of a machine is called **anthropometry**. For a discussion of the topic, see Module B Section 10.

Fatigue contributes significantly to the causes of accidents, and designers should aim to reduce this to a minimum by considering:

- the correct placing of controls
- the positioning of operating stations and height of work tables
- the provision of seating
- suitable access to the work station

TYPES OF MACHINERY GUARD

FIXED GUARD: A fixed guard should be fitted wherever practicable and should, by design, prevent access to dangerous parts of machinery. It should be robust to withstand the process and environmental conditions. Its effectiveness will be determined by the method of fixing and the size of any openings, allowing for an adequate distance between the opening and the danger point. This may be determined by national standards or Regulations. The guard should only be removable with the use of a special tool, such as a spanner or wrench.

INTERLOCKED GUARD: The essential principles of an interlocked guard are that the machine cannot operate until the guard is closed, and the guard cannot be opened until the dangerous parts of the machine have come fully to rest. Interlocked guards can be mechanical, electrical, hydraulic, pneumatic or a combination of these. Interlocked guards and their components have to be designed so that any failure of them does not expose people to danger.

CONTROL GUARD: If the motion of the machine can be stopped quickly, control guards can be used. The principle of control guarding is that the machine must not be able to operate until the guard is closed. When the guard is closed by the operator, the machine's operating sequence is started. If the guard is opened, the machine's motion is stopped or reversed, so the machine must come to rest or reverse its motion quickly for this technique to be effective. Generally, the guard in a control system is not locked closed during the operation of the machine.

AUTOMATIC GUARD: An automatic guard operates by physically removing from the danger area any part of the body exposed to danger. It can only be used where there is adequate time to do this without causing injury, which limits its use to slow-moving machinery.

DISTANCE GUARD: A distance guard prevents any part of the body from reaching a danger area. It could take the form of a fixed barrier or fence designed to prevent access.

ADJUSTABLE GUARD: Where it is impracticable to prevent access to dangerous parts (they may be unavoidably exposed during use), adjustable guards (fixed guards with adjustable elements) can be used. The amount of protection given by these guards relies heavily on close supervision of the operative, correct adjustment of the guard and its adequate maintenance.

SELF-ADJUSTING GUARD: This guard is automatically opened by the movement of the workpiece and returns to its closed position when the operation is completed.

TRIP DEVICES: These automatically stop or reverse the machine before the operative reaches the danger point. They rely upon sensitive trip mechanisms and on the machine being able to stop quickly (which may be assisted by a brake). Examples include trip wires and mats containing switches which stop the machine when they are trodden on.

TWO-HAND CONTROL DEVICES: These devices force the operator to use both hands to operate the machine controls. However, they only provide protection for the operator, and not for anyone else who may be near the danger point. Guards should be arranged to protect all persons. Where these devices are provided the controls should be spaced well apart and/or shrouded, the machine should only operate when both controls are activated together, and the control system should require resetting between each cycle of the machine.

SELECTION OF SAFEGUARDS

Fixed guards provide the highest standard of protection, and should be used where practicable where access to the danger area is not required during normal operation. The following gives guidance on the selection of safeguards (in order of merit):

a) Where access to the danger area is not required during normal operation:
   1. Fixed guard, where practicable
   2. Distance guard
   3. Trip device
b) Where access to the danger area is required during normal operation:
   1. Interlocking guard
   2. Automatic guard
   3. Trip device
   4. Adjustable guard
   5. Self-adjusting guard
   6. Two-hand control

Safe Machinery Design and Guarding

LEGAL REQUIREMENTS

Manufacturers, designers, importers and suppliers have a duty to ensure that their equipment is safe so far as is reasonably practicable (Health and Safety at Work etc. Act 1974, Section 6). A detailed extension of these duties is also contained in the Management of Health and Safety at Work Regulations 1992. Manufacturers and others in the supply chain have further duties to comply with the specific safety requirements set out in minimum safety standards for equipment manufactured, designed, imported and supplied in the European Community.

Every employer must ensure, so far as is reasonably practicable, the health and safety of his employees and others (Health and Safety at Work etc. Act 1974, Sections 2(1) and 3)). This extends to the provision and maintenance of equipment, plant and systems of work by the employer, that are safe (Section 2 (2)(a)). This obligation covers machinery guarding and layout.

More specific guarding (fencing) requirements are contained in the Provision and Use of Work Equipment Regulations 1992, the Factories Act 1961 and the Offices, Shops and Railway Premises Act 1963. For a full discussion of the requirements of these Regulations and their relationships see Module D Sections 7, 11 and 12. For an example of Regulations relating specifically to specific processes or industries see Module D Section 17 – the Abrasive Wheels Regulations 1970. Much of this specific type legislation is revoked under the Provision and Use of Work Equipment Regulations 1992 (see Module D Section 7).

SELF-ASSESSMENT QUESTIONS

1. Select a machine you are familiar with, and identify the ways in which the controls have been designed to prevent danger.

2. In what circumstances would plastic sheeting not be an adequate material to use in machine guarding?

REVISION

Danger can arise from machinery hazards – traps, entanglement and non-machinery hazards – the operator, noise

Safety can be achieved by design in the following ways:

- Intrinsic safety
- Control selection
- Failure to safety
- Maintenance and isolation procedures
- Safety by position
- Machine layout

Five dangers associated with machines:

- Traps
- Impact
- Contact
- Entanglement
- Ejection

Nine types of guard:

- Fixed
- Interlocked
- Control
- Automatic
- Distance
- Adjustable
- Self-adjusting
- Trip
- Two-hand control

INTRODUCTION

Mechanical handling techniques have improved efficiency and safety, but have introduced other sources of potential injury into the workplace. Cranes, powered industrial trucks and fork lifts, and conveyors are the primary means for mechanical handling. In all circumstances the safety of the equipment can be affected by the safety of operating conditions, workplace hazards and the operator.

CRANES

Basic safety principles for all mechanical equipment apply to cranes. These principles are that the equipment should be of good construction, made from sound material, of adequate strength and be free from obvious faults. All equipment should be tested and regularly examined to ensure its integrity. The equipment should always be properly used.

WHAT CAN GO WRONG

OVERTURNING can be caused by weak support, operating outside the machine's capabilities and by striking obstructions

OVERLOADING by exceeding the operating capacity or operating radii, or by failure of safety devices

COLLISION with other cranes, overhead cables or structures

FAILURE of **support** – placing over cellars and drains, outriggers not extended, made-up or not solid ground, or of **structural components** of the crane itself

OPERATOR errors from impaired/restricted visibility, poor eyesight, inadequate training

LOSS OF LOAD from failure of lifting tackle or slinging procedure

HAZARD ELIMINATION

Matters which require attention to ensure the safe operation of a crane include:

IDENTIFICATION AND TESTING: every crane should be tested and a certificate should be issued by the seller and following each test. Each should be identified for reference purposes, and the safe working load clearly marked. This should never be exceeded, except under test conditions.

MAINTENANCE: cranes should be inspected regularly, with any faults repaired immediately. Records of checks and inspections should be kept.

SAFETY MEASURES: a number of safety measures should be incorporated for the safe operation of the crane. These include:

a) Load Indicators – of two types
   i. Load/radius indicator
   ii. Automatic safe load indicator, providing audible and visual warning
b) Controls – should be clearly identified and of the "dead-man" type
c) Overtravel switches – limit switches to prevent the hook or sheave block being wound up to the cable drum
d) Access – safe access should be provided for the operator and for use during inspection and maintenance/emergency
e) Operating position – should provide clear visibility of hook and load, with the controls easily reached
f) Passengers – should not be carried without authorisation, and never on lifting tackle
g) Lifting tackle – chains, slings, wire ropes, eyebolts and shackles should be tested/examined, be free from damage and knots as appropriate, be clearly marked for identification and safe working load, and be properly used (no use at or near sharp edges, or at incorrect sling angles

OPERATING AREA: all nearby hazards, including overhead cables and bared power supply conductors, should be identified and removed or covered by safe working procedures such as locking-off and permit systems. Solid support should be available and on new installations the dimensions and strength of support required should be specified. The possibility of striking other cranes or structures should be examined.

OPERATOR TRAINING: crane operators and slingers should be fit and strong enough for the work. Training should be provided for the safe operation of the particular equipment.

Mechanical Handling

POWERED INDUSTRIAL TRUCKS

Trucks should be of good construction, free from defects and suitable for the purpose in terms of capacity, size and type. The type of power supply to be used should be checked, because the nature of the work area may require one kind of power source rather than another. In unventilated confined spaces, internal combustion engines will not be acceptable because of the toxic gases they produce. Trucks should be maintained so as to prevent failure of vital parts, including brakes, steering and lifting components. Any damage should be reported and corrected immediately. Overhead protective guards should be fitted for the protection of the operator. Powered industrial trucks and their attachments should only be operated in accordance with the manufacturer's instructions.

WHAT CAN GO WRONG

OVERTURNING from manoeuvring with load elevated, driving at too high a speed, sudden braking, striking obstructions, use of forward tilt with load elevated, driving down a ramp with the load in front of the truck, turning on or crossing ramps at an angle, shifting loads, unsuitable road or support conditions

OVERLOADING by exceeding the maximum lifting capacity of the truck

COLLISION with structural elements, pipes, stacks or with other vehicles

FLOOR FAILURE because of uneven or unsound floors, or by exceeding the load capacity of the floor. The capacity of floors other than the ground level should always be checked before using trucks on them

LOSS OF LOAD can occur if devices are not fitted to stop loads slipping from forks

EXPLOSIONS AND FIRE may arise from electrical shorting, leaking fuel pipes, dust accumulation (internal combustion) and from hydrogen generation during battery charging. The truck itself can be the source of ignition if operated in flammable atmospheres

PASSENGERS should not be carried unless seats and other facilities are provided for them.

HAZARD ELIMINATION

Matters which require attention for the safe operation of powered industrial trucks include:

OPERATING AREA: the floor should be of suitable construction for the use of trucks. It should be flat and unobstructed where movement of machines is expected, with gullies and openings covered. Storage and stacking areas should be properly laid out with removal of blind corners. Passing places need to be provided where trucks and people are likely to pass each other in restricted spaces, and routes need to be clearly defined with adequate visibility. Pedestrians should be excluded from operating areas, or alternatively clearly defined gangways should be provided, with trucks given priority. Suitable warning signs will be required to indicate priorities.

Lighting should be adequate to facilitate access and stacking operations. Loading bays should be appropriately designed and stable with chocks provided to place behind or beneath wheels. Ramps and slopes should not exceed 1:10 unless the manufacturer specifies that use on steeper gradients under load is acceptable. Battery charging areas should be separate, well-ventilated and lit with no smoking or naked lights permitted. Battery lifting facilities should be provided. Reversing lights should be fitted, where possible, especially where pedestrians may share floor space with trucks.

TRAINING should be provided for operators in the safe operation of their equipment, followed by certification.

CONVEYORS

The most common types of conveyor are belt, roller or screw conveyors.

WHAT CAN GO WRONG

TRAPPING – limbs can be drawn into in-running nips

CONTACT – with moving parts such as drive elements, screw conveyors

ENTANGLEMENT – with rollers, drive mechanisms

STRIKING – materials falling from heights, incorrectly handled

HAZARD ELIMINATION

BELT CONVEYORS: require guards or enclosures at the drums, which are the main hazard because of the presence of trapping points between belt and drum. These are also required where additional trapping points or nips occur as the belt changes direction or at guide plates or feed points. Guards may be required along the length of a conveyor in the form of enclosures or trip wires to cut off supply. Safe access at appropriate intervals should be provided over long conveyor runs.

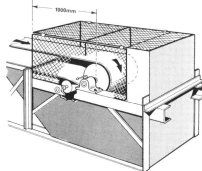

ROLLER CONVEYORS: where rollers may be either power-driven or free-running, guards at power drives are required. Other hazards can be present, which also require guarding – these include areas where in-running nips are created, where alternate rollers are power driven or when a belt is fitted. Walkways should be provided if access is required over the conveyor mechanism.

SCREW CONVEYORS: should be guarded to prevent access at all times. Repairs and maintenance should only

be undertaken when the drive is locked off.

LEGAL REQUIREMENTS

Mechanical handling is covered by the general duties provisions of the Health and Safety at Work etc. Act 1974, and the more detailed requirements of the Management of Health and Safety at Work Regulations 1992. The Provision and Use of Work Equipment Regulations 1992 apply to all equipment first taken into use after January 1st 1993. See Module D Section 7 for a short description of transitional arrangements for older equipment. For factory premises, Sections 26 and 27 of the Factories Act 1961 apply to cranes and lifting gear alongside all the requirements of the Provision and Use of Work Equipment Regulations 1992. On construction sites, the Construction (Lifting Operations) Regulations apply to some mechanical handling operations in building, construction and civil engineering work alongside Regulations 5, 6, 9, 14, 20, 23, and 24 of the Provision and Use of Work Equipment Regulations 1992.

SELF-ASSESSMENT QUESTIONS

1. What checks should be made before a crane is used in a workplace?

2. A powered lift truck overturns and the operator is injured. List the potential causes of the overturning.

REVISION

Cranes fail through:

- Overturning
- Overloading
- Collision
- Foundation failure
- Structural failure
- Operator error
- Loss of load
- Lack of maintenance

Powered lift trucks fail through:

- Overturning
- Overloading
- Collision
- Floor failure
- Loss of load
- Operator error
- Presence of passengers

Conveyor hazards:

- Trapping
- Contact
- Entanglement
- Non-machinery hazards, including noise, vibration
- Striking by objects

Manual Handling

INTRODUCTION

It has long been recognised that the manual handling of loads at work contributes significantly to the number of workplace injuries with approximately a quarter of all reported accidents attributed to these activities. The majority of these injuries result in more than three days' absence from work, almost half of the total being back injuries.

Generally, about half of the resulting injuries are sprains or strains of the lower back, with other types of injury including cuts, bruises, fractures and amputations. Many of the injuries are of a cumulative nature rather than being attributable to any single handling incident.

INJURIES RESULTING FROM MANUAL HANDLING

Some of the more common types of injury resulting from manual handling are considered here. It is important to remember that the spine is undergoing a process of "aging" which will affect the integrity of the spine. For example, loss of "spinal architecture" (i.e. the natural curvature of the spine) produces excessive pressure on the edges of the discs wearing them out faster. These effects are not considered here.

Disc injuries – 90% of back troubles are attributable to **disc lesions**. When stood upright, forces are exerted directly through the whole length of the spine and in this position it can withstand considerable stress. However, when the spine is bent, most of the stress is exerted on only one part (usually at the part where the bending occurs). Also due to the bending of the spine in one place, all the stress is exerted on one side of the intervertebral disc, thus "pinching" it between the vertebrae. This 'pinching effect' may scar and wear the outer surface of the disc so that at some time it becomes weak and, under pressure, ruptures. It is also important to remember that the discs dry out with age, making them less flexible and functional, and more prone to injury. The disc contents are highly irritating to the surrounding parts of the body, causing an inflammatory response. This is known as a **prolapsed interverte-bral disc lesion**.

Ligament/tendon injuries – Ligaments and tendons are connective tissues, and hold the back together. Ligaments are the gristly straps that bind the bones together whilst tendons attach muscles to other body parts, usually bones. Repetitive motion of the tendons may cause inflammation. Both can be pulled and torn, resulting in sprains. Any factor that produces tightness in ligaments and tendons predisposes the back to sprains (such as age effects and cold weather). Two main ligaments run all the way down the spine to support the vertebrae.

Muscular/nerve injuries – The muscles in the back form long, thick bands that run down each side of the spine. They are very strong and active, but tiring of these muscles can result in aches and pains, and can create stress on the discs. Postural deformities can result from damage to the muscles. Fibrositis (rheumatic pain) can also result. Nerves can be become trapped between the elements of the spine causing severe pain and injury.

Hernias – A hernia is a protrusion of an internal organ through a gap in a wall of the cavity in which it is contained. For example, any compression of the abdominal contents towards the naturally weak areas, may result in a loop of intestine being forced into one of the gaps or weak areas formed during the development of the body. So, when the body is bent forward, possibly during a lift, the abdominal cavity decreases in size causing a compression of internal components and increasing the risk of hernias.

Fractures, abrasions and cuts – These can result from dropping the objects that are being handled (possibly because of muscular fatigue), falling whilst carrying objects (perhaps as a result of poor house-keeping), from other inadequacies in the working environment such as poor lighting, or from the contents of the load.

INJURY DURING MANUAL HANDLING

The injuries highlighted above can result from the lifting, pushing, pulling and carrying an object during manual handling.

Lifting – Compressive forces on the spine, its ligaments and tendons can result in some of the injuries identified above. High compressive forces in the spine can result from lifting too much, poor posture and incorrect lifting technique. Prolonged compressive stress causes what is known as "creep-effect" on the spine, squeezing and stiffening it. If the spine is twisted or bent sideways when lifting, the added tension in the ligaments and muscles when the spine is rotated considerably increases the total stress on the spine to a dangerous extent.

Pushing and pulling – Stresses are generally higher for pushing than pulling. Because the abdominal muscles are active as well as the back muscles, the reactive compressive force on the spine can be even higher than when lifting. Pushing also loads the shoulders and the rib-cage is stiffened, making breathing more difficult.

Carrying – Carrying involves some static muscular work which can be tiring for the muscles, the back, shoulder, arm and hand depending on how the load is supported. A weight held in front of the body induces more spinal stress than one carried on the back. Likewise, a given weight held in one hand is more likely to cause fatigue than if it were divided into equal amounts in each hand. As with pushing, which loads the shoulders, carrying objects in front of the body or on the shoulders may restrict the rib-cage. Thus the way in which a load is carried makes a great difference to the fatiguing effects.

MANUAL HANDLING ASSESSMENTS

The Manual Handling Operations Regulations 1992 (See Module D Section 8) establish a clear hierarchy of measures to reduce the risk of injury when performing manual handling tasks. To summarise, manual handling operations which present a risk must be avoided so far as is reasonably practicable; if these tasks cannot be avoided, then each task must be assessed and as a result of that assessment, the risk of injury must be reduced for each particular task identified so far as is reasonably practicable.

It is important to remember that what is required initially is not a full assessment of each of the tasks, but an "appraisal" of those manual handling operations which involve a risk that cannot be dismissed as trivial, to determine if they can be avoided. Consideration of a series of questions will be useful in completing this stage of the exercise. These include:

Is there a risk of injury? – An understanding of the types of potential injury will be supplemented with the past experiences of the employer, including accident/ill-health information relating to manual handling and the general numerical guidelines contained in official guidance.

Is it reasonably practicable to avoid moving the load? – This will be useful in establishing "authorised" manual handling tasks within a certain workplace or department. Some work is dependent on the manual handling of loads and cannot be avoided (as, for example, in refuse collection). If it is reasonably practicable to avoid moving the load then the initial exercise is complete and further review will only be required if conditions change.

Is it reasonably practicable to automate or mechanise the operation? – Introduction of these measures can create different risks (introduction of fork-lifts creates a series of new risks) which require consideration.

The aim of the **full assessment** is to evaluate the risk associated with a particular task and identify control measures which can be implemented to remove or reduce the risk (possibly mechanisation and/or training). For varied work (such as done in the course of maintenance, construction or agriculture) it will not be possible to assess every single instance of manual handling. In these circumstances, each type or category of manual handling operation should be identified and the associated risk assessed. Assessment should also extend to cover those employees who carry out manual handling operations away from the employers premises (such as delivery drivers).

The assessment must be kept up to date. It needs reviewing whenever it may become invalid, such as when the working conditions or the personnel carrying out those operations have changed. Review will also be required if there is a significant change in the manual handling operation, which may, for example, affect the nature of the task or the load.

Before beginning an assessment, the views of staff and employees can be of particular use in identifying manual handling problems. Involvement in the assessment process should be encouraged, particularly in reporting problems associated with particular tasks.

Records of accidents and ill-health are also valuable indicators of risk as are absentee records, poor productivity and morale, and excessive product damage. Individual industries and sectors have also produced information identifying risks associated with manual handling operations.

Schedule 1 to the Manual Handling Operations Regulations specifies four inter-related factors of which the assessment should take account. The answers to the questions specified about the factors form the basis of an appropriate assessment. These are:

The task

Do they involve:
- holding or manipulating loads at distance from the trunk?
- unsatisfactory bodily movement or posture, especially –
    twisting the trunk?
    stooping?
    reaching upwards?
- excessive movement of loads, especially:
    excessive lifting or lowering distances?
    excessive carrying distances?
- excessive pushing or pulling of loads?
- risk of sudden movement of loads?

Manual Handling

- frequent or prolonged physical effort?
- insufficient rest or recovery periods?
- a rate of work imposed by a process?

The loads

Are they:
- heavy?
- bulky or unwieldy?
- difficult to grasp?
- unstable, or with contents likely to shift?
- sharp, hot, or otherwise potentially damaging?

The working environment

Are there:
- space constraints preventing good posture?
- uneven, slippery or unstable floors?
- variations in levels of floors or work surfaces?
- extremes of temperature or humidity?
- conditions causing ventilation problems or gusts of wind?
- poor lighting conditions?

Individual capability

Does the job:
- require unusual strength, height, etc?
- create a hazard to those who might reasonably be considered to be pregnant or to have a health problem?
- require special information or training for its safe performance?

Other factors

- is movement or posture hindered by PPE or clothing?

REDUCING THE RISK OF INJURY

In considering the most appropriate controls, an ergonomic approach to 'designing' the manual handling operation will optimise the health, safety and productivity associated with the task. The task, the load, the working environment, individual capability and the inter-relationship between these factors are all important elements in deciding optimum controls designed to fit the operation to the individual rather than the other way around. Techniques of risk reduction include:

- **Mechanical assistance** This involves the use of handling aids of which there are many examples. One could be the use of a lever, which would reduce the force required to move a load. A hoist can support the weight of a load whilst a trolley can reduce the effort needed to move a load horizontally. Chutes are a convenient means of using gravity to move loads from one place to another.

- **Improvements in the task** Changes in the layout of the task can reduce the risk of injury by, for example, improving the flow of materials or products. Improvements which will permit the use of the body more efficiently, especially if they permit the load to be held closer to the body, will also reduce the risk of injury. Improving the work routine by reducing the frequency or duration of handling tasks will also have a beneficial effect. Using teams of people and personal protective equipment (PPE) will also contribute to a reduced risk of injury. All equipment supplied for use during handling operations (e.g. handling aids and PPE) should be maintained and there should be a defect reporting and correction system.

- **Reducing the risk of injury from the load** The load may be made lighter by using smaller packages/containers or specifying lower packaging weights. Additionally, the load may be made smaller, easier to manage, easier to grasp (for example, by the provision of handles), more stable and less damaging to hold (clean, free from sharp edges etc.).

- **Improvements in the working environment** By removing space constraints, improving the condition and nature of floors, reducing work to a single level, avoiding extremes in temperature and excessive humidity and ensuring that adequate lighting is provided.

- **Individual selection** The health, fitness and strength of an individual can clearly affect ability to perform manual handling tasks. Health screening is an important tool in selection for manual handling tasks. Knowledge and training have important roles to play in reducing the number of injuries resulting from manual handling operations.

MANUAL HANDLING TRAINING

A training programme should include mention of:

- Dangers of careless and unskilled handling methods
- Principles of levers and the laws of motion
- Functions of the spine and muscular system
- Effects of lifting, pushing, pulling and carrying, with emphasis on harmful posture
- Use of mechanical handling aids
- Selection of suitable clothing for lifting and necessary protective equipment

- Techniques of
  a) identifying slip/trip hazards
  b) assessing weight of loads and how much can be handled by the individual without assistance
  c) bending the knees, keeping the load close to the body when lifting (but avoiding tension and knee-bending at too sharp an angle)
  d) breathing, and avoiding twisting and sideways bending during exertion
  e) using the legs to get close to the load, making best use of body and load weight

There are several guiding principles for the safe lifting of weights. These are:

- Secure grip
- Proper foot position

- Bent knees and comfortably straight back
- Keep arms close to the body
- Keep chin tucked in
- Body weight used to advantage

(All attending should have an opportunity to practise under supervision)

Other factors which should be discussed with trainees include:

- Personal limitations (age, strength, fitness, girth)
- Nature of loads likely to be lifted (weight, size, rigidity)
- Position of loads
- Working conditions to minimise physical strain
- The requirements of the Manual Handling Operations Regulations 1992 and risk assessments

SELF-ASSESSMENT QUESTIONS

1. What factors should be considered by management before lifting of weights is authorised?

2. Identify areas in your workplace where mechanised handling techniques could be used instead of manual handling.

REVISION

Five types of injury:

- Disc injuries
- Ligament/tendon injuries
- Muscular/nerve injuries
- Hernias
- Fractures, abrasion, cuts

Training programmes should include:

- The dangers of bad lifting technique
- The principles of leverage
- The functions of the body in lifting
- Demonstration of good technique
- Use of mechanical aids
- Opportunity to practise
- Requirements of the Regulations and risk assessments

Eight points for safe lifting:

- Check load characteristics – weight, size, position, destination
- Be aware of personal limitations, ask for assistance if necessary
- Take secure grip
- Keep back straight and knees bent
- Keep arms close to body
- Keep the chin tucked in
- Be aware of body weight and how to use it to advantage
- Co-ordinate two or more persons handling an object

Four factors in assessment:

- Task
- Load
- Working environment
- Individual capability

Access Equipment

INTRODUCTION

Many work and home injuries result from failures or falls involving access equipment which has been incorrectly selected, erected, used or maintained. Access equipment is frequently used for short duration and emergency work without full consideration of a safe method of work. Each task should be assessed, and a suitable means of access chosen based upon an evaluation of the **work** to be done, the **duration** of the task, the **working environment** (and its constraints), and the **capability** of the person or people carrying out the task.

There are many different types of access equipment. This Section covers general principles, and the following:

- Ladders, stepladders and trestles
- General access scaffolds
- Scaffold towers
- Suspended cradles and chairs
- Mast-elevated work platforms
- Power-operated work platforms

Other, highly-specialised, equipment is available, and the general principles will apply to their use. Usually, they have been specially designed for particular tasks and manufacturers' information should be used in operator training.

GENERAL PRINCIPLES

Accidents using access equipment occur because one or more of the following common problems have not been controlled in advance, or was thought to be an acceptable risk under the circumstances. They are caused by:

- Faulty design of the access structure itself
- Inappropriate selection where safer alternatives could have been used
- Subsidence or failure of base support
- Structural failure of suspension system
- Structural failure of components
- Structural failure through overloading
- Structural failure through poor erection/inspection/maintenance
- Structural failure through overbalancing
- Instability through misuse or misunderstanding
- Over-reaching and overbalancing
- Climbing while carrying loads
- Slippery footing – wrong footwear, failure to clean
- Falls from working platforms and in transit
- Unauthorised alterations and use

- Contact with obstructions and structural elements
- Electrical and hydraulic equipment failures
- Trapping by moving parts

LADDERS, STEPLADDERS AND TRESTLES

The key points to be observed when selecting and using this equipment are:

LADDERS

1. See whether an alternative means of access is more suitable. Take into account the nature of work and duration, the height to be worked at, what reaching movements may be required, what equipment and materials may be required at height, the angle of placement and the foot room behind rungs, and the construction and type of ladder.

2. Check visually whether the ladder is in good condition and free from slippery substances.

3. Check facilities available for securing against slipping

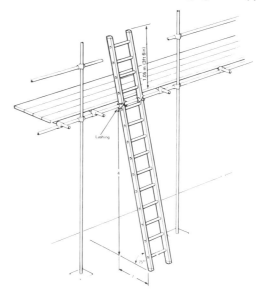

– tie at top, secure at bottom or footed by a second person if no more than 5m (15ft) height access is required.

4. Ensure the rung at the step-off point is level with the working platform or other access point, and that the ladder rises at least 1m (3'6") above this point, unless there is a separate handhold.

5. A landing point for rest purposes is required every 9m (30ft).

6. The correct angle of rest is approximately 75 degrees (corresponds to a ratio of one unit horizontally at the foot for every four units vertically).

7. Stiles (upright sections) should be evenly and adequately supported.

8. Ladders should be inspected regularly and not painted, except for identification.

9. Ladders not capable of repair should be destroyed.

10. Metal ladders (and wooden ladders when wet) are conductors of electricity and should not be placed near or carried beneath low power lines.

STEPLADDERS

1. Stepladders are not designed to accept side loading.

2. Chains or ropes to prevent overspreading are required, or other fittings designed to achieve the same result. Parts should be fully extended.

3. Stepladders should be levelled for stability on a firm base.

4. Work should not be carried out from the top step.

5. Over-reaching should be avoided by moving the stepladder – if this is not possible, another method of access should be considered.

6. Equipment should be maintained free from defects, and not painted except for identification marking. Regular inspection is required.

7. No more than one person should use a stepladder at one time.

TRESTLES

1. Trestles are suitable for short duration work only, and as board supports.

2. They should be free from defects and inspected regularly.

3. Trestles should be levelled for stability on a firm base.

4. Platforms based on trestles should be fully boarded, adequately supported and provided with edge protection where appropriate.

5. Safe means of access should be provided to trestle platforms, usually by a stepladder.

GENERAL ACCESS SCAFFOLDS

There are three main types of access scaffold commonly constructed from steel tubing or available in commercial patented sections. These are:

1. Independent tied scaffolds, which are temporary structures independent of the structure to which access is required but tied to it for stability.

2. Putlog scaffolds, which rely upon the building (usually under construction) to provide structural support to the temporary scaffold structure through an arrangement of putlog tubes placed into the wall.

3. Birdcage scaffolds, which are independent structures normally erected for interior work which have a large area and normally only a single working platform.

The key points to be observed when specifying, erecting and using scaffolds are:

1. Select the correct design with adequate load-bearing capacity.

2. Ensure adequate foundations are available for the loads to be imposed.

3. The structural elements of the scaffold should be provided and maintained in good condition.

4. Structures should be erected by competent persons or under the close supervision of a competent person, in accordance with any design provided and with applicable local or national Regulations or Codes.

5. All working platforms should be fully boarded, with adequate edge protection, including handrails or other means of fall protection, nets, brickguards and/or toeboards to prevent materials or people falling from the platforms.

6. All materials resting on platforms should be safely stacked, with no overloading.

7. Adequate and safe means of access should be provided to working platforms.

8. Unauthorised alterations of the completed structure should be prohibited.

9. Inspections of the structure are required, first upon completion and then at appropriate intervals afterwards, usually weekly. Details of the results should be entered into an inspection register or logbook.

Access Equipment

SCAFFOLD TOWERS

Scaffold towers are available commercially in forms comparatively easy to construct. They may also be erected from traditional steel tubing and couplers. In either form, competent and trained personnel are required to ensure that all necessary components are present and in the right place. Many accidents have occurred because of poor erection standards; a further common cause is overturning.

The key points to be observed in the safe use of scaffold towers are:

1. Erection should be in accordance with the manufacturer's or supplier's recommendations.

2. Erection should be carried out by experienced, competent persons.

3. Towers should be stood on a firm level base, with wheel castors locked if present.

4. Scaffold equipment should be in good condition, free from patent defects including bent or twisted sections, and properly maintained.

5. The structure should be braced in all planes, to distribute loads correctly and prevent twisting and collapse.

6. The ratio of the minimum base dimension to the height of the working platform should not exceed 1:3 in external use, and 1:3.5 in internal use, unless the tower is secured to another permanent structure at all times. Base ratios can be increased by the use of outriggers, but these should be fully extended and capable of taking loads imposed at all times.

7. Free-standing towers should not be used above 9.75m (30ft) unless tied. The maximum height when tied should not exceed 12m (40ft).

8. A safe means of access should be provided on the narrowest side of the tower. This can be by vertical ladder attached internally, by internal stairways, or by ladder sections designed to form part of the frame members. It is not acceptable to climb frame members not designed for the purpose.

9. Trapdoors should be provided in working platforms where internal access is provided.

10. Platforms should be properly supported and fully boarded.

11. Guardrails, toeboards and other appropriate means should be provided to prevent falls of workers and/or materials.

12. Mobile scaffold towers should never be moved while people are still on the platform. This is a significant cause of accidents.

13. Ladders or stepladders should not be placed on the tower platform to gain extra height for working.

SUSPENDED ACCESS (CRADLES AND BOATSWAIN'S CHAIRS)

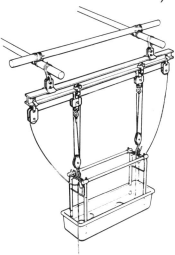

A **suspended access system** includes a working platform or cradle, equipped with the means of raising or lowering itself when suspended from a roof rig.

A **boatswain's (or bosun's) chair** is a seating arrangement provided with a means of raising or lowering with a suspension system. This should only be used for very short duration work, or in positions where access by other means is impossible.

The key points to be observed in the safe installation and use of this equipment are:

1. It should be capable of taking the loads likely to be imposed on it.

2. Experienced erectors only should be used for the installation.

3. Supervisors and operators should be trained in the safe use of the equipment, and in emergency procedures.

4. Inspections and maintenance are to be carried out regularly.

5. Suspension arrangements should be installed as designed and calculated.

6. All safety equipment, including brakes and stops, should be operational.

7. The marked safe working load should not be exceeded, and wind effects should also be considered.

8. Platforms should be free from obstruction, and fitted with edge protection.

9. The electrical supply is not to be capable of inadvertent isolation, and should be properly maintained.

10. Adverse weather conditions should be defined so that supervisors and operators know what is not considered acceptable.

11. All defects noted are to be reported and rectified before further use of the equipment.

12. Safe access is required for the operators, and unauthorised access is to be prevented.

13. Necessary protective measures for those working below and the public should be in place before work begins.

ELEVATED WORK PLATFORMS

Generally, this equipment consists of three elements:

- Mast(s) or tower(s) which support(s) a platform or cage
- A platform capable of supporting persons and/or equipment
- A chassis supporting the tower or mast.

The key points to be observed in the erection and use of this equipment are:

1. Only trained personnel should erect, operate or dismantle the equipment.

2. The manufacturer's instructions on inspection, maintenance and servicing should be followed.

3. Firm, level surfaces should be provided, and outriggers are to be extended before use or testing, if provided.

4. Repairs and adjustments should only be carried out by qualified people.

5. The safe working load of the equipment should be clearly marked on it, be readily visible to the operator, and never be exceeded.

6. Raising and lowering sequences should only be initiated if adequate clearance is available.

7. The platform should be protected with edge guardrails, toeboards and provided with adequate means of access.

8. Emergency systems should only be used for that purpose, and not for operational reasons.

9. Unauthorised access into the work area should be prevented using ground barriers.

10. Contact with overhead power cables should be prevented, by preliminary site inspection and by not approaching closer than a given distance. This distance can be obtained in respect of the particular power lines, from the power supply company, where necessary.

POWER-OPERATED MOBILE WORK PLATFORMS

A wide variety of equipment falls into this category, ranging from small mobile tower structures with self-elevating facilities, to large vehicle-mounted hydraulically-operated platforms.

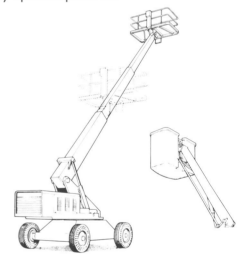

The key points to be observed in their use are:

1. Operator controls should be at the platform level, with override at ground level for emergencies.

2. There should be a levelling device fitted to the chassis to ensure verticality in use.

3. Supervision should prevent use of the equipment during adverse weather conditions.

4. Outriggers, where provided for increased stability, should be fully extended and locked into position before the equipment is used/raised, in accordance with the manufacturer's instructions. The wheels may also require locking.

5. Materials and/or persons should not be transferred to and from the platform while in the raised position.

Access Equipment

6. Training is required for operators before they are allowed to use the equipment in field conditions unsupervised.

7. When fitted, scissor mechanisms require the provision of adequate fixed guards, so as to prevent trapping of the operator or others in the scissor mechanism during raising or lowering.

8. The equipment requires regular inspection, servicing, maintenance and testing in accordance with the manufacturer's instructions.

LEGAL REQUIREMENTS

Requirements for safe access and equipment can be found in the Health and Safety at Work etc. Act 1974 and the more detailed requirements of the Management of Health and Safety at Work Regulations 1992 and The Workplace (Health, Safety and Welfare) Regulations 1992. The most complete set of requirements is contained within the Construction (Working Places) Regulations 1966. These should be followed as being standards normally achievable "so far as is reasonably practicable" in other industries. There is much guidance material available to amplify more general legal duties, a selection of which can be found in the Selected References Section at the end of this Module.

SELF-ASSESSMENT QUESTIONS

1. Complete the boxes in the table below, identifying the common failure modes associated with each type of access equipment.

2. A scaffold collapses as a result of overloading. What are the ways in which this could have been prevented?

Self Assessment Question 1	Ladders, Stepladders, Trestles	General Access Scaffolds	Scaffold Towers	Suspended Cradles and Chairs	Mast-elevated Work Platforms	Power-operated Work Platforms
Faulty design of the access structure itself						
Inappropriate selection (safer alternatives available)						
Subsidence or failure of foundations/footings						
Structural failure of suspension system						
Structural failure of component strength						
Structural failure through overloading						
Structural failure (poor erection/inspection/maintenance)						
Structural failure through overbalancing						
Instability through misuse or misunderstanding						
Over-reaching and overbalancing						
Climbing carrying loads						
Slippery footing – wrong footwear, failure to clean						
Falls from working platforms and in transit						
Unauthorised alterations and use						
Contact with obstructions and structural elements						
Electrical and hydraulic equipment failures						
Trapping by moving parts						

REVISION

Access equipment should be selected after consideration of:

- The work to be done
- The duration of the work
- The location of the work
- Means of access to the work area
- The environment
- The people who will do the work
- The people who will erect the access equipment
- Available technical specifications and information

Common causes of accidents using access equipment:

- Design fault
- Not using safer alternative
- Unsound base support
- Structural failure
- Instability
- Over-reaching
- Contact with obstructions and energy sources
- Untrained personnel
- Poor maintenance
- Incorrect erection and/or use

Transport Safety

INTRODUCTION

The safe operation of vehicles results from planning and activity, not chance. In the majority of vehicle accidents, the principal factors are driver failure and vehicle failure, both of which can be controlled. A relatively small proportion of accidents is due to vehicle mechanical failure.

CAUSES OF TRANSPORT ACCIDENTS

Transport accidents occur because of:

- Contact – with structures, services
- Overturning – through incorrect loading, speeding, surface conditions
- Collision – with other vehicles or pedestrians
- Impact – materials falling or the vehicle overturning onto the operator
- Entanglement – in dangerous parts of machinery or controls
- Explosion – when charging batteries or inflating tyres
- Operator/supervisor error – through inadequate training or experience

PREVENTING TRANSPORT ACCIDENTS

These accidents are preventable by the use of a planned approach involving:

- Driver selection, training, certification and supervision – the use of vehicles by unauthorised persons must be prevented. Authorisation should depend upon the individual's progress through a training programme on the specific type(s) of vehicle to be driven, and selection of the individual for the task.
- Control of visiting drivers – any local rules must be communicated to visiting drivers, either in writing, as a contract condition, or by the provision of suitable traffic safety signs or markers.
- Traffic control in the workplace – includes making decisions on needs, priorities, right of way and the separation of pedestrian and motor traffic.
- Accident investigation – including reporting system, subsequent analysis of reports, and corrective action follow-up.
- Maintenance procedures – safety, economy and efficiency all benefit from periodic vehicle checks and inspections, in addition to local or national requirements of Regulations or Codes.

DRIVER SELECTION, TRAINING AND SUPERVISION

Selection will require an evaluation of age, experience, driving record and attitude. People who drive safely also have other qualities such as courtesy and ability to get on well with others. These can be used to pick potential safe drivers. Local or national driver qualification requirements must also be observed.

Training of drivers may be for remedial, refresher or special reasons in addition to basic instruction for drivers unfamiliar with the vehicle to be driven. Generally, driver training courses should cover applicable local and national driving rules, company driving rules, what to do in the event of an accident to comply with local, national and company requirements, and defensive driving techniques.

Supervision must insure that all vehicle drivers engage in safe practices. They should also be aware of their drivers' safety performance and how this compares with other company areas and with comparable industry figures. They will also be responsible for investigation and recording of accidents.

CONTROL OF VISITING DRIVERS

For any workplace, visitors are unlikely to be familiar with work practices, layout and local rules and will require to have this information presented to them in an appropriate way. This can be done by publication of written material, or by positioning appropriate signs. Control of visitors must be exercised, since if the breaking of local rules is condoned this will have a negative effect on attitudes of those who observe them.

TRAFFIC CONTROL IN THE WORKPLACE

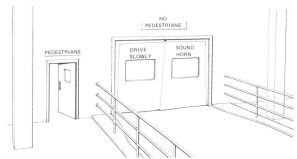

Adequate clearance should be provided for the safe movement of vehicles, identifying overhead or floor-level obstructions. A combination of design and good housekeeping will be needed to identify and control these hazards. Vehicles and pedestrians should be separated where practicable, with warning lights and/or signs displayed. Safe crossing places should be clearly marked, and fitted with mirrors to improve visibility into blind areas. Vehicle speeds should be controlled with speed limits, backed where necessary with speed ramps. Movements should be properly supervised when reversing or access is difficult or blind, using recognised signals. Access routes should be lit where practicable, especially where pedestrians and vehicles share routes.

Loading and unloading of vehicles should take place in designated areas without obstruction to other traffic. Vehicles not in use or broken down should be left where obstruction is minimised. Vehicle loads must be both stable and secure. Drivers should be adequately protected against falling objects or roll-over. Vehicle keys should be removed when not in use, to immobilise and guard against theft or misuse.

The vehicles should be checked to ensure that dangerous parts of machinery are guarded, and that there is safe means of access and egress. Reminders of the correct means of access, and the need to wear appropriate clothing, should be part of the training of drivers and maintenance staff.

TRANSPORT ACCIDENT INVESTIGATION

In addition to the requirements of local or national laws and Codes concerning reporting and recording of accidents, feedback of information gained from investigation is very useful. Information of this kind will have consequences for the driver training programme, trigger discussion with the persons involved, add to company experience and records, and provide a means of assessing each driver.

Each driver should be required to complete a standard report form for each accident involving a vehicle. Personal investigation by supervisory staff should also be made where possible, to verify the driver's data and obtain any necessary extra information.

A record should be maintained for each driver, as well as for each vehicle. Driver records can form the basis for awards, and review of them will identify repetitions of inappropriate behaviour which may be rectified by further training.

MAINTENANCE PROCEDURES

Planned maintenance procedures prevent accidents and delays due to mechanical failures, minimise repair downtime and prevent excessive wear and breakdown. Drivers and operators are usually the first to notice when defects develop, and should check their vehicles against a basic checklist before work starts. Vehicle checklists should cover the following items:

- Brakes
- Headlights
- Stoplights and indicators
- Tyres
- Screens and wipers
- Steering wheel
- Glass
- Horn
- Mirrors
- Instruments
- Exhaust system
- Emergency equipment
- Ignition problems
- Connecting cables

In addition to other health and safety considerations, **vehicle repair** requires attention to the following:

Brakes must be applied and wheels chocked, especially before entry under vehicle bodies. Raised bodies must be propped unless fitted with proprietary devices for stability in the raised position. Axle stands must be used in conjunction with jacks, and hydraulic jacks should not be relied on alone. When charging batteries, the risk of explosion or burns should be eliminated. Precautions are also required to prevent explosion risks when draining, repairing or carrying out hot work near fuel tanks. Tyre cages should be provided for inflating vehicle tyres. Steps must be taken to eliminate exposure to dangerous dusts and fumes, for example asbestos (brakes) and operating internal combustion engines in confined areas. Only trained personnel should carry out maintenance work.

CARRYING HAZARDOUS LOADS

Manufacturers of hazardous substances should have final responsibility for safety, in the sense that they are likely to have most knowledge of the specific properties of the substances they make, and to be best able to recommend safe handling and emergency procedures. This information should be provided in data sheets (at least) and be made available to all concerned, including drivers and dispatch points.

Transport Safety

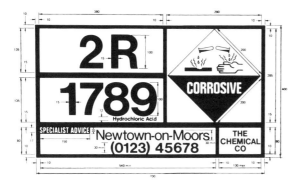

There are many Codes and Regulations covering this subject. They contain specific, detailed requirements about containers and packaging, and their marking. They also have requirements on identifying substances by their properties, including flammability, and ways in which information has to be given to those handling and carrying the loads and to the public. This is done in a number of ways, including by the display of special signs on the packaging and on the vehicle. Requirements differ slightly between countries, and there are also international Standards; reference to these and to national Codes and Regulations is essential if loads are carried between countries.

The instruction and training of drivers is also covered extensively by Codes and Regulations. The provisions usually include the need to alert emergency services in the event of accidents, use of any emergency equipment required to be carried on the vehicle, and the steps which can be taken safely following an accident to minimise the consequences before the arrival of emergency services. The driver should also be able to give these services the information they need to know about the load, including methods of treating spillages and/or fire. Special attention is required to the selection of mature drivers for hazardous loads, with appropriate skills and abilities.

LEGAL REQUIREMENTS

Generally, the Health and Safety at Work etc. Act 1974 and the more detailed requirements of the Management of Health and Safety at Work Regulations 1992 will apply to all aspects of transport safety. The carrying of hazardous loads is covered in detail in legal provisions. In a work of this length it is not possible to do more than indicate the enormous number of Regulations, Codes and Standards which can apply nationally, and internationally, to the movement of transport and goods of various kinds.

SELF-ASSESSMENT QUESTIONS

1. A vehicle maintenance fitter insists he is not exposed to danger when carrying out his work. Identify the hazards to which he is exposed.

2. The elements of a planned approach to transport safety are listed here. For your workplace, make notes about any which require further attention, planning or control.

Driver selection
Driver training
Driver supervision
Control of visiting drivers
Workplace traffic control
Accident investigation
Maintenance procedures

REVISION

Transport accidents occur through:

- Contact
- Overturning
- Collision
- Impact
- Entanglement
- Explosion
- Operator/supervisor error
- Poor maintenance

The **Five** main elements of transport safety are:

- Driver selection, instruction and supervision
- Control of visiting drivers
- Workplace traffic control
- Accident investigation
- Preventative maintenance and vehicle repair

Classification of Dangerous Substances

INTRODUCTION

Dangerous substances can be categorised in many ways: when discussing or describing them in health and safety terms they are categorised according to the type of harm they can cause. Many can fall into more than one category. Some can cause harm after a single exposure or incident, others may have only long-term effects on the body following repeated exposure.

CLASSIFICATIONS

Generally, substances can be placed into one or more of the following broad categories:

EXPLOSIVE OR FLAMMABLE – dangerous because of their potential to release energy rapidly, or because the product(s) of the explosion or combustion are harmful in other ways. Examples of these are organic solvents.

HARMFUL – substances which if inhaled, ingested or enter the body through the skin present a limited risk to health.

IRRITANTS – adversely affect the skin or respiratory tract, for example acrylates. Some people overreact to irritants like isocyanates which are **sensitizers** and can cause allergic reactions.

CORROSIVES – substances which will attack chemically either materials or people, for example strong acids and bases.

TOXICS – substances which prevent or interfere with body functions in a variety of ways. They may overload organs such as the liver or kidneys. Examples are chlorinated solvents and heavy metals, such as lead.

CARCINOGENS, MUTAGENS and TERATOGENS – prevent the correct development and growth of body cells. Carcinogens cause or promote the development of unwanted cells as cancer. Teratogens cause abnormal development of the embryo, producing stillbirth or birth defects. Mutagens alter cell development and cause changes in future generations.

AGENTS OF ANOXIA – vapours or gases which reduce the oxygen available in the air, or prevent the body using it effectively. Carbon dioxide, carbon monoxide and hydrogen cyanide are examples.

NARCOTICS – produce dependency, and act as depressors of brain functions. Organic solvents are common narcotics.

OXIDISING – a substance which gives rise to an exothermic reaction (gives off heat) when in contact with other substances, particularly flammable ones.

LABELLING

In most jurisdictions, substances with potential for harm are required to be safely packaged and labelled according to various Codes of Practice and Regulations. These involve the use of descriptive labels showing the hazard in the form of pictograms, accompanied by required descriptive phrases which convey necessary information in a shortened form. The labels also carry the details of the manufacturer. In all cases, management should not be content to rely upon the packaging to provide sufficient information to make decisions about safe storage, use, handling, transportation and disposal of the substance.

LEGAL REQUIREMENTS

Regulations covering labelling and packaging of substances often contain definitions and classifications. This short Section gives a summary of the categories usually found in Regulations, and directs attention to standards and guidance at the end of the Module.

EXPLOSIVE   HIGHLY FLAMMABLE   IRRITANT   CORROSIVE   TOXIC   HARMFUL   OXIDIZING

Classification of Dangerous Substances

SELF-ASSESSMENT QUESTIONS

1. Find and fill in examples of each of the categories of classification for substances in your own workplace. If none are present for a category, consult a reference book and find a new example apart from those mentioned in the text.

   EXPLOSIVE / FLAMMABLE
   HARMFUL
   IRRITANT
   CORROSIVE

   TOXIC
   CARCINOGENS / MUTAGENS / TERATOGENS
   AGENTS OF ANOXIA
   NARCOTIC
   OXIDISING

2. Can you think of other, simple, ways in which substances could be classified in your workplace so as to give an indication of their potential for harm? Are there any drawbacks to your classification?

REVISION

Substances are categorised according to the type of harm they can cause.

Nine classifications:

- Explosive/Flammable
- Harmful
- Irritant
- Corrosive
- Toxic
- Carcinogens/Mutagens/Teratogens
- Agents of anoxia
- Narcotic
- Oxidising

INTRODUCTION

Chemical incidents can affect many people including operators, nearby workers, others on site and members of the public. Systems for the control and safe use of chemicals and substances are crucial to the safe operation of any workplace which handles, uses, stores, transports or disposes of them, regardless of the quantity involved. Planning for safety requires the setting out of broad principles which can then be translated into specific detail at the workplace.

PLANNING FOR CHEMICAL SAFETY

The way in which we plan for chemical safety is based upon the principles for the control of all types of risk, which will be familiar to you by the time you complete working through this book. Section 3 of Module C discusses the principles of risk assessment, which are applicable to chemical safety. The stages of control include risk assessment as the basis for control measures. As this is the first time we use it, the six stages of control are summarised here:

- Identification of the hazard to be controlled
- Assessment of the risk
- Control of the risk
- Training of staff
- Monitoring the effectiveness of the strategy
- Necessary record-keeping

HAZARD DATA SHEET

Triethylamine
$(C_2H_5)_3N$

Other names
—

Description
Colourless liquid with a strong ammonia-like odour.

Threshold Limit Value
25ppm

Flash point
−17°C (1°F)

Boiling point
90°C (194°F)

Lower explosive limit
1.2%

Hazards
1 Highly flammable vapour.
2 Contact with eyes and skin causes burns.

Precautions
1 No smoking, naked flames, hot elements or other ignition sources.
2 If flammable concentrations are likely, use flameproof electrical fittings.
3 Ensure adequate earthing.
4 Avoid exposure to vapour at levels above the Threshold Limit Value.
5 Use gloves, eye protection and protective clothing at all times when handling this material.

Storage
1 Store in steel drums, tightly sealed and clearly labelled.
2 Drums should be stored in a flammable liquids store, which should be a cool, dry, well-ventilated place, protected from direct sunlight, heat and frost.
3 No smoking, naked flames, hot elements or other ignition sources.

Fire fighting procedure
Extinguish with foam, carbon dioxide or dry powder.

First aid measures

Inhalation
Remove to fresh air.
In severe cases, or if exposure has been great, obtain medical attention.

Skin contact
Wash with plenty of water.
Remove grossly contaminated clothing and wash before re-use.
In severe cases obtain medical attention.

Eye contact
Irrigate with plenty of water.
Obtain medical attention.

Ingestion
Obtain immediate medical attention.

Spillage
1 Eliminate all possible sources of ignition.
2 Wear goggles or face shield, gloves and protective clothing.
3 Use breathing apparatus in handling anything other than a small spillage.
4 Absorb on sand, earth or proprietary absorbent material and dispose of as chemical waste.
5 Wash site of spillage thoroughly with water.

EEC warning phrases
Highly flammable.
Irritating to eyes.
Irritating to respiratory system.
Keep away from sources of ignition — no smoking.
In case of contact with eyes, rinse immediately with plenty of water and seek medical advice.
Do not empty into drains.

Relevant legislation includes:
1 Disposal of Poisonous Wastes Act 1974
2 Control of Pollution Act 1974.
3 Highly Flammable Liquids and Liquefied Petroleum Gases Regulations 1972.

IDENTIFICATION OF THE HAZARD: In the present case, we already know the hazard, as "chemical safety" is the objective. "**Hazard**" means the inherent property or ability of something to cause harm – it is not the same as "**risk**", which is the likelihood of the harm to be caused in given circumstances. The extent of the hazard posed by a chemical or substance can be established by the use of information – from the supplier, the manufacturer, from records, historical knowledge and other sources of information. Some substances are labelled on their containers; others, such as reaction products, are not brought into the workplace from outside, have no containers and must be identified and checked against reference material.

ASSESSMENT OF THE RISK: The risk presented by a chemical or substance will not be dependent only upon its physical and chemical properties, but will also be a function of the way it is used, where it is used, and how it may be misused or mishandled. Assessment of these factors is essential in determining local risk at a particular location.

CONTROLLING THE RISK: This can be achieved by elimination of the hazardous substance entirely, by its substitution with a less hazardous alternative, mechanical/remote handling, total enclosure, exhaust ventilation, special techniques such as wet methods or use of personal protective equipment. These control techniques may be used in combination, but they are set in order of effectiveness and desirability. There is also the possibility of reducing exposure time and numbers of personnel at risk.

TRAINING OF OPERATORS: Instruction in the nature of the hazards they face and the degree of risk, associated with the procedures devised to eliminate hazards where possible, and the effective control of the remaining risks is necessary. Especially, this will ensure that operators do not use their own solutions to problems. In-house procedures involving chemicals are often written down for all to see and understand; where this is done it is important that the procedures are publicised, available and readable. Language difficulties with minority groups should be resolved and translations provided where necessary.

MONITORING EFFECTIVENESS: Regular monitoring of the effectiveness of control measures is needed to ensure that change – in work conditions, quality, information, and systems – is taken into account and anticipated in procedures. This can include regular monitoring of work practices, sampling of air/process quality and purchasing procedures. Except for workplaces where little change is expected, monitoring and appropriate measuring should take place at regular intervals specified in the procedures. This is especially important where personal protective equipment is relied upon as a major method of controlling risks.

RECORD-KEEPING: Keeping records helps to ensure the regular maintenance of plant at prescribed intervals, and

Chemical Safety

permits awareness of change. Medical surveillance of operators may be required by local or national Codes and Regulations.

RISK CONTROL

The following topics should be considered and adopted as appropriate to reduce exposure to risks:

PLANT DESIGN: Chemical processes can be controlled by introducing containment and process control – the safe design of plant processes and the use of automated and/or computer-controlled systems. Plant should not be altered without specific authority and review.

SAFE SYSTEMS OF WORK: These include the use of detailed operating and emergency instructions, permit-to-work systems (especially for maintenance operations), and installation, use and maintenance of protective equipment.

TRANSPORT: Controlled movement of chemicals in containers or vehicles intended for that purpose reduces the likelihood of exposure to operators and third parties. Adequate emergency procedures will be required.

STORAGE: Chemicals should not be affected by adjacent processes or chemical storage. External storage of chemicals should be the method of choice, as this gives quick dispersion, minimises sources of ignition and provides secure storage which reduces the risk of damage to containers. Features of storage facilities should include a hard standing, bund retention, adequate separation, adequate access for mechanical handling, contents gauges for fixed tanks, relief valves if required, security of stacking, fire fighting facilities and warning signs in accordance with applicable Codes and Regulations.

WASTE DISPOSAL: Adequate steps must be taken to ensure the correct disposal of waste, preventing harm to members of the public and damage to the environment, in accordance with legislative requirements.

EMERGENCY PROCEDURES: These must be drawn up for production and storage areas, and for the site as a whole. Emergency services and operatives alike should be briefed on the procedures to be followed, and adequate time must be given for training and rehearsal at regular intervals.

The relative importance of each of these topics will depend upon the assessed degree of risk at the workplace.

LEGAL REQUIREMENTS

The general duties Sections of the Health and Safety at Work etc. Act 1974 apply to chemical safety in the same way that they apply to all work activities, and provide protection for those affected by work activities as well as those at work. More detailed requirements are contained in the Management of Health and Safety at Work Regulations 1992. In addition, specific Regulations controlling substances hazardous to health apply to all work activities (the COSHH Regulations, also covered in Module D), and work sites handling, storing or using certain chemicals in quantity are covered by other Regulations which are not discussed here in detail (the CIMAH – or Control of Industrial Major Hazards – Regulations 1984). Broadly, the latter requires notification of the presence of more than specified quantities of specified chemicals and the submission of information on control measures in the form of a report.

Presence on site of more than 25 tonnes of "dangerous substances" as defined by the Dangerous Substances (Notification and Marking of Sites) Regulations 1990 triggers notification procedures and marking requirements involving the use of specified safety signs. Disposal of all waste is controlled by a number of Acts and Regulations, which should be consulted.

SELF-ASSESSMENT QUESTIONS

1. Is operator training sufficient by itself to prevent chemical accidents?

2. Give an example of a substance with different degrees of risk when used and when stored.

REVISION

There are **six** stages of control:

- Hazard identification
- Risk assessment
- Risk control
- Staff training
- Monitoring
- Record-keeping

Hazard – the inherent ability to cause harm

Risk – likelihood of harm in particular circumstances

There are **six** topics for control of chemical risk:

- Plant design
- Systems of work
- Transport
- Storage
- Disposal
- Emergencies

Electricity and Electrical Equipment

INTRODUCTION

The ratio of fatalities to injuries is higher for electrical accidents than for most other categories of injury – if an electrical accident occurs, the chances of a fatality are about one in forty. The consequences of contact with electricity are: electric shock, where the injury results from the flow of electricity through the body's nerves, muscles and organs and causes abnormal function to occur (the heart stops, for example); electrical burns resulting from the heating effect of the current which burns body tissue; and electrical fires caused by overheating or arcing apparatus in contact with a fuel (See the next Section).

CAUSES OF ELECTRICAL FAILURES

Failures and interruption of electrical supply are most commonly caused by:

- Damaged insulation
- Inadequate systems of work
- Inadequate overcurrent protection (fuses, circuit breakers)
- Inadequate earthing
- Carelessness and complacency
- Overheated apparatus
- Earth leakage currents
- Loose contacts and connectors
- Inadequate ratings of circuit components
- Unprotected connectors
- Poor maintenance and testing

PREVENTING ELECTRICAL FAILURES

These failures can be prevented by regular attention to the following points:

Earthing – providing a suitable electrode connection to earth through metal enclosure, conduit, frame etc. Regular inspections and tests of systems should be carried out by a competent person.

System of work – when working with electrical circuits and apparatus, the supply should be switched and locked off; apparatus should be checked personally by the worker to ensure that it is "dead"; permit-to-work systems should be used in high-risk situations

previously identified. Working on live circuits and apparatus should only be permitted under circumstances which are strictly controlled and justified in each case. Rubber or other non-conducting protective equipment may be required. Barriers and warning notices should be used to ensure that persons not involved in the work directly do not expose themselves to risk. The use of insulated tools and equipment will always be necessary.

Insulation – where work is required near uninsulated parts of circuits. In all circumstances, making the apparatus "dead" must be considered as a primary aim, and rejected only if the demands of the work make this not practicable. A variety of permanent or temporary insulators may be used, such as cable sheathing and rubber mats.

Fuses – these are strips of metal placed in circuits which melt as a result of overheating in the circuit, effectively cutting off the supply. Different fuses "melt" at predetermined current flows. A factor in the selection of fuses is that there is a variable and appreciable delay in their action, which may expose those at risk to uninterrupted current for unacceptable periods.

Circuit breakers – detect electromagnetically and automatically any excess current flow and cut off the supply to the circuit.

Residual current devices – detect earth faults and cut off the supply to the circuit.

Competency – only properly trained and suitably experienced people should be employed to install, maintain, test and examine electrical circuits and apparatus.

STATIC ELECTRICITY

Electrical charging can occur during the movement of powders and liquids, which can cause sparking, and this may ignite a dust cloud or a flammable vapour. Less seriously, static electricity in other working environments may be a cause of annoyance to workers, and be a factor in other types of accident where attention has been distracted by static discharge.

Protection against static electricity can be achieved by earthing and not using or installing equipment which can become statically charged (See the next Section). Operators should also wear anti-static shoes.

ELECTRICAL EQUIPMENT

Electrical tools used should be selected and operated bearing in mind the following considerations:

Substitution – electrical tools and equipment may be replaced with pneumatic equipment – which have their own dangers!

Switching off circuits and apparatus – this must be readily and safely achievable.

Reducing the voltage – use of the lowest practicable voltage should be practised in every circuit.

Cable and socket protection – should be provided to protect against physical and environmental effects which could have adverse consequences on the integrity of circuits and apparatus, such as rain.

Plugs and sockets – must be of the correct type and specification, and meet local and national Regulations and Codes of Practice.

Maintenance and testing – should be carried out at regular and prescribed intervals by competent and experienced personnel. Recording of the results and values measured will provide a baseline figure for assessing any subsequent deterioration in performance or quality.

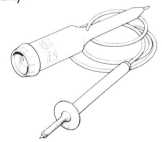

Explosive atmospheres – require the careful consideration and selection of equipment for service in dusty or flammable environments. The type of equipment to be used is commonly specified by law.

THE TREATMENT OF ELECTRIC SHOCK

> **NOTE**: *Advice given in this Section is not intended as a substitute for or alternative to attending a recognised first-aid course.*

RECOGNISING A CASUALTY

Apart from proximity to a source of electrical energy, casualties may show symptoms of asphyxia and have no discernible pulse. Violent muscular contraction caused by contact with high voltage supplies can throw casualties some way from the original point of contact. In cases of electric shock breathing and heart beat can stop together, accounting for the pallid blue tinge to the skin sometimes seen. There may also be contact burns visible which can indicate contact with electricity.

BREAKING THE CONTACT

Low voltage supplies, commonly at 240 volts, can be switched off at the mains or meter if readily accessible. If not, or it cannot be found quickly, the supply can be turned off at the plug point or by pulling out the plug. The important point to remember when removing the casualty from the source is not to become a casualty in turn. Dry insulating material such as newspaper or wood can be used to push or pull the casualty clear by wrapping around or pushing at arms and legs near the source. Do not touch the casualty directly.

Higher voltage supplies such as those found in overhead or underground power lines are frequently fatal, and produce deep burns. It is important not to approach a casualty touching or lying close to one of these lines, until the line has been isolated by the supply authority. Closer approach than this can lead on occasions to electrical arcing. At higher voltage levels, insulating material requires special properties not possessed by anything likely to be close to hand.

TREATMENT

The order of priority for treatment is:

1. Resuscitation if required

Electricity and Electrical Equipment

2. Place in Recovery Position if casualty is still unconscious but breathing normally
3. Treat any burns
4. Treat for shock
5. Remove to hospital in all cases where resuscitation was required, casualty was unconscious, casualty received burns or developed symptoms of shock. Pass to hospital any information on duration of electrical contact.

RESUSCITATION

Mouth-to-mouth ventilation should be used where a casualty is not breathing, and ventilation of the lungs is required to restart breathing reflexes. **Cardio-pulmonary resuscitation** (CPR) will be needed if the heart has stopped in addition to breathing. Both these techniques require practice; descriptions of them are beyond the scope of this Section. Details can be found in first-aid manuals, and they are taught on first-aid courses.

RECOVERY POSITION

Putting a casualty into this position maintains an open airway and improves stability of the body. Placing of the casualty into the Recovery Position is not difficult, but a description of the method is beyond the scope of this Section. The position to be achieved is one where the casualty lies on the front and side, with head supported for comfort and slightly flexed backwards to maintain the airway. The uppermost hand is near the face with the arm bent at the elbow to give support, and the uppermost leg is bent at the hip and knee forming a right angle at the knee with the thigh well forward.

BURNS TREATMENT

For severe burns likely to be associated with electric shock, it is important to immobilise the burned part, and to cover the burned area with a dry sterile unmedicated dressing or anything similar, and secure with a bandage if available. The important things NOT TO DO are:

1. Do not allow clothing to restrict any swelling of the part
2. Do not apply anything to the burn
3. Do not break blisters or attempt to treat or touch the injured area
4. Do not remove anything sticking to the burned part.

Sips of cold water will help the casualty to replace lost fluid, but give no medication.

SHOCK TREATMENT

Traumatic shock can prove fatal even when other injuries have been treated. It is a consequence of a reduction in volume of body fluids. The casualty feels faint, sick and possibly thirsty, and has shallow breathing or yawning. The pulse rate may increase but may become weak and irregular. The skin is pale and cold.

Treatment involves keeping the casualty warm, loosening tight clothing and elevating the legs if possible. Do not give anything by mouth, apply external heat, or allow the casualty to smoke. Avoid moving the casualty unless this cannot be avoided.

LEGAL REQUIREMENTS

These are now covered completely by the Electricity at Work Regulations 1989, which are discussed in Module D and amplify the general duties placed on employers, employees and the self-employed by the Health and Safety at Work etc. Act 1974. More detailed requirements are contained in the Management of Health and Safety at Work Regulations 1992. Arrangements and provisions for the treatment of electric shock are contained in the Health and Safety (First-Aid) Regulations 1981 (See Module D, Section 18).

SELF-ASSESSMENT QUESTIONS

1. What preventative measures can be taken against electrical failure?

2. An electrical shock occurs as a result of using a drill outside. What factors might have contributed to this accident?

REVISION

Three results of electrical failure:

- Electric shock
- Electrical burns
- Electrical fires

Ten preventative measures:

- Earthing
- Safe system of work
- Insulation
- Fuses
- Circuit breakers
- Residual current devices
- Competency
- Reduced voltage
- Maintenance
- Isolation

Fire

INTRODUCTION

The control of the start and spread of fire is an important feature of the prevention of accidents and damage. Fire needs fuel, oxygen and a source of energy to ignite it; these are known as the "fire elements". Examples of them are wood, air and heat in the right combination. The ratio of fuel to oxygen is crucial; too much or too little of either will not permit a fire to start. Also, the source of ignition must be above a certain energy level.

There are five main hazards produced by fire; oxygen depletion, flame/heat, smoke, gaseous combustion products and structural failure of buildings.

CLASSIFICATION OF FIRE

There are four main categories, based upon the fuel and the means of extinction. These are:

CLASS A: Fires which involve solid materials, predominantly of an organic kind, forming glowing embers. Examples are wood, paper and coal. The extinguishing mode is by cooling, and is achieved by the use of water.

CLASS B: Fires which involve liquids or liquefiable solids; they are further subdivided into:

CLASS B1, which involve liquids soluble in water, for example methanol. They can be extinguished by carbon dioxide, dry powder, water spray, light water and vaporising liquids.

CLASS B2, which involve liquids not soluble in water, such as petrol and oil. They can be extinguished by foam, carbon dioxide, dry powder, light water and vaporising liquids.

CLASS C: Fires which involve gases or liquefied gases resulting from leaks or spillage, e.g methane or butane. Extinguishment can be achieved by using foam or dry powder in conjunction with water to cool any leaking container involved.

CLASS D: Fires which involve metals such as aluminium or magnesium. Special dry powder extinguishers are required to fight these, which may contain powdered graphite or talc. No other extinguisher type should be used.

Electrical fires, which involve the electricity supply to live equipment, can be dealt with by extinguishing mediums such as carbon dioxide, dry powder or vaporising liquids, but not water. Electricity is a cause of fire, not a category of fire. Electrical fires have recently been removed from the traditional 'categories' of fire, for this reason.

Knowledge of the correct type of extinguisher to use, or install, in areas at particular risk is essential.

FIRE PROTECTION

There are three strategies for protecting against and dealing with fire:

Structural design precautions – providing protection through insulation, integrity and stability permitting people to escape, for example compartmentalisation, smoke control, unobstructed means of escape

Fire detectors and alarms – activated by sensing heat, flame, smoke or flammable gas, for example heat detectors, radiation detectors, smoke detectors

Fire fighting – with portable extinguishers or fixed fire fighting equipment (manually or automatically operated)

STRUCTURAL AND DESIGN PRECAUTIONS

Controlling the likelihood of fire occurring can be achieved by the removal of one of the fire elements by:

a) **Removing sources of ignition** Heat is the most common source of ignition, and can result from:
   - Heating systems – hot water pipes, which can be lagged
   - Friction – parts of a machine, which require proper lubrication
   - Hot surfaces – machine panels, to be cooled by design, insulation or maintenance
   - Electricity – arcing or heat transfer, removed by selection or design of appropriate electrical protection
   - Static electricity – through relative motion and/or separation of two different materials, prevented by providing an earth path, humid atmosphere, high voltage device or ion exchange
   - Smoking – spent smoking materials, prevented by restricting the activity or confining it to defined areas

Heat transfer can be by radiation, conduction or convection.

b) **Controlling the fuel** Supply of materials should be kept to a minimum during work or in storage areas. The main fuel sources are:
   - Waste, debris and spillage
   - Gases & vaporising liquids
   - Flammable liquids
   - Flammable compressed gases and liquefied gases
   - Liquid sprays
   - Dusts

MEANS OF FIRE DETECTION AND ALARMS

Fire, or flammable atmospheres, can be detected in the following ways:

Heat detection – sensors operate by the melting of a metal (fusion detectors) or expansion of a solid, liquid or gas (thermal expansion detectors)

Radiation detection – photoelectric cells detect the emission of infra-red/ultra-violet radiation from the fire

Smoke detection – using ionising radiations, light scatter (smoke scatters a beam of light), obscuration (smoke entering a detector prevents light from reaching a photoelectric cell)

Flammable gas detection – measures the amount of flammable gas in the atmosphere and compares the value with a reference value

Fire alarms must make a distinctive sound, audible in all parts of the workplace. The meaning of the alarm sound must be understood by all. They may be manually or automatically operated.

FIRE FIGHTING

Fires can be extinguished by suppression or removal of one of the three fire elements. This can be done by fuel starvation, by preventing fuel flow or removing its source and by containment. Covering a fire will prevent oxygen from reaching fuel, and cooling with water removes a primary source of ignition. Interference with the combustion process can also be done as a chemical process.

A fire plan should be available and practised regularly in even the smallest premises. Such a plan will be of value to professional fire fighters, who need to know the location of flammable and hazardous material in premises. The associated evacuation process should also be practised at intervals.

SIGNS AND TRAINING

Signs should be provided to identify courses of action to be taken, means of escape and location of fire fighting equipment. Training should be given in the use of systems and equipment provided, and to minimise panic.

MEANS OF ESCAPE IN CASE OF FIRE

The purpose of a means of escape is to enable people confronted by fire to proceed in the opposite direction to an exit away from the fire, through a storey exit, a protected staircase and/or a final exit to ground level in open air away from the building. Any means of escape must not rely on rescue facilities from the fire service.

Means of escape in buildings other than those with only a ground floor generally consist of three distinct areas. These are:

- any point on a floor to a staircase
- the route down a staircase
- the route from the foot of the staircase to the open air, clear of the building

Essentially, there are two areas associated with means of escape:

- the area in which people escaping from a fire are at some risk (the unprotected zone)
- areas where risk is reduced to an acceptable minimum (protected zones). Protected zones at an exit must be fully protected, using walls or partitions which have a fire-resisting capability.

DETERMINING THE MEANS OF ESCAPE

Types of means of escape will be determined by:

- Occupancy characteristics – numbers, physical capability of users
- Building uses – residential commercial, manufacturing, entertainment etc.
- Construction characteristics of the building
- Evacuation times
- Occupant movements

TRAVEL DISTANCE

This is a technical term used in fire safety, and is the greatest allowable distance which has to be travelled from the fire to reach the beginning of a protected zone of a means of escape or final exit. This distance must be limited, although it will vary according to the fire risk. The calculation of travel distance involves consideration of a number of factors, and only guidelines are given in publications on the subject. The most comprehensive of these is in the Home Office publication "Fire Precautions Act 1971: guide to fire precautions in existing places of work that require a fire certificate", ISBN 0 11 340906 0.

PROTECTED LOBBIES AND STAIRCASES

As parts of the protected zone these must be of fire-resisting capability. Doors must be self-closing. The

Fire

objective is to ensure that people entering lobbies and staircases on any floor are to be able to remain within their safe confines until reaching the ground. Such protection serves the purposes of preventing smoke and heat from obstructing the staircase so making it unsuitable for escape purposes, and of preventing a fire from spreading into a staircase from one storey and passing to another.

Staircases should be protected, usually by a ventilated lobby. Staircases should be:

- Continuous
- Have risers and treads of dimensions specified in Standards
- Be ventilated
- Have any internal glazing and spaces underneath able to resist fire
- Be fitted with handrails on both sides

LIFTS AND ESCALATORS

Lifts and escalators must be disregarded as potential means of escape. Lifts have limited capacity, may be delayed in arriving and have potential for mechanical failure. Escalators have restricted width, and the stair pitch when not in operation may lead to possible injury, congestion and panic.

WIDTH AND CAPACITY OF ESCAPE ROUTES

There should be no risk of overcrowding, delay or formation of "bottlenecks" in the event of mass evacuation.

FIRE DOORS

These have the main functions of preserving the safe means of escape by retarding the passage of fire (so they must be fire-resistant) and retarding the spread of smoke and hot gases in the initial stages of a fire.

FINAL EXIT DOORS

General principles of design and function apply to all means of escape final exit doors, which may be modified in particular cases. The operating principles are that these doors must:

a) open in the direction of escape
b) preferably not open onto steps
c) not be revolving doors
d) be capable of being opened at any time whilst the building is occupied, preferably without using keys
e) be protected from fires in the basement or in adjacent buildings
f) be of adequate width
g) be provided with adequate dispersal space.

OTHER MEANS OF EXIT

Means of escape via roof – generally unacceptable, but if it must be used then it must be protected from smoke, fire and falls. A safe means of access must be provided to the ground.

High level access – across roof or other areas at height to protected means of escape. If required to be a means of escape it must be protected against fire, be clearly defined and protected against falls.

Ramps – ramps may be used if they have a non-slip surface, a uniform pitch, do not exceed a slope of 1:10 and are provided with handrails.

External fire escapes – these are not desirable, as they are subject to weather conditions such as ice and snow which makes them potentially dangerous. However, they are sometimes necessary. They must be protected against fire, and windows adjacent to them must be fire-resistant and not open outwards so as to obstruct passage down the fire escape.

ACTION TO BE TAKEN IN THE EVENT OF FIRE

In the presence of fire, panic and the urge to get away are natural reactions. Information about the action to take, and where possible practice in that action, are essential to ensure the optimum response in the event of fire.

NOTICES TO EMPLOYEES AND OTHERS

Copies of notices giving simple guidance on what to do in the event of fire should be displayed in all workplaces and premises where persons could be at risk from fire. The details of each notice, which may be written as a safety sign, will vary and depend upon the layout, fire risk and circumstances likely to be encountered. For this reason they should be specially written. Diagrams of layouts showing the means of escape in each case will be helpful.

In drafting notices, attention should be paid to the needs of those who are infrequent visitors to the premises and who may have had no need or opportunity to practise any fire drill there. Instructions in the notice should therefore be straightforward and simple to understand. They may need to be given in translated form if a number of non-English speakers are likely to be present. This will be the case in hotels, for example.

Notices should explain simply what to do if a fire is discovered, or if the fire alarm is sounded. In the latter case, the actual sound made by the fire alarm should be explained if there is any possibility of confusion.

Fire instructions
If you discover a fire
1. Immediately operate the nearest fire alarm call point;
2. Attack the fire, if possible, with the appliances provided but without putting yourself at risk;

On hearing the fire alarm
3. _____ will call the Fire Brigade immediately;
4. Leave the building and report to the person in charge of the assembly point at: _____
5. The warden or deputy on the affected floor or department will take charge of any evacuation and ensure that no one is left in the area.

Use the nearest available exit

Do not use lifts
Do not stop to collect personal belongings
Do not re-enter the building

Instructions for Calling the Fire Brigade. For Display at all Telephone Exchanges and all Night Extension Telephones

NOTICE CONTENTS

The displayed notices should cover the following points:

1. Identification of fire alarm sound
2. What to do when the alarm is sounded (e.g. close windows, switch off appropriate equipment, leave the room, close doors behind, follow signs, report to assembly point at specified location, do not re-enter building until advised)
3. What to do if a fire is discovered (e.g. how to raise the alarm, what fires may be tackled by staff, evacuation procedure previously described).

LEGAL REQUIREMENTS

Major provisions are contained in the Fire Precautions Act 1971, as amended by the Fire Safety and Safety of Places of Sport Act 1987, and in detailed provisions concerning fire certification. It is anticipated that there will be some changes to this legislation in the near future as a result of the EC Directive relating to health and safety requirements in the workplace (this Directive brought about the Workplace (Health, Safety and Welfare) Regulations 1992). However, consultation on the mechanics of implementing the fire requirements of the Directive for workplaces is still continuing at the time of writing.

The reader will be aware that the general duties of the Health and Safety at Work etc. Act 1974, amplified in detail by the Management of Health and Safety at Work Regulations 1992, apply to the taking of fire precautions in general terms. However, the latter Regulations require employers to identify circumstances where situations presenting serious and imminent danger to employees and others, in **all** workplaces. These must be identified in risk assessments, and the procedures written down. Details must be given to employees and others who may be exposed to those situations, and fire (and possibly bomb) risks will be the only ones that need to be covered to comply with the requirement (Regulation 7). It is clear that virtually all employers must publish detailed emergency fire procedures.

Fire

SELF-ASSESSMENT QUESTIONS

1. How would you handle a fire involving a flammable gas cylinder?

2. List the design features handling fire protection in your work area.

REVISION

Three fire elements:

- Fuel
- Oxygen
- Ignition

Four classes of fire:

- Class A – wood, paper
- Class B – liquid
- Class C – gases
- Class D – metals (but **not** electrical fires, extinguishable by all mediums except water)

Three protection strategies:

- Structural design precautions
- Fire detectors & alarms
- Fire fighting

Five fire hazards:

- Oxygen depletion
- Flame/heat
- Smoke
- Gaseous combustion products
- Structural failure

Construction Safety

INTRODUCTION

Over the last 150 years the construction industry has had an unenviable safety record. The Victorians instituted a massive programme of building and civil engineering with little thought for the safety of the huge workforce – as a result thousands were killed. We live now in more enlightened times, where the taking of risks at work and the exposure of non-employees to risk are seen as less acceptable than before. However, significant organisational changes in the industry itself have had consequences for accident prevention, the effects of which have balanced out improved attitudes.

Traditionally, a contractor would be engaged directly by a client in consultation with specialist advisers such as architects. The contractor would employ the majority of workers on a site full-time, and specialist subcontractors being hired in by him or directly by the client. Those who were self-employed would be taken on to make up numbers or to provide specialist skills.

Changes in taxation rules amongst many other social and economic factors have increased the numbers of self-employed on sites to the extent that the majority of workers in the industry now regard themselves as such, although whether the Health and Safety Executive, Inland Revenue or personal injury lawyer would agree is another matter. The contractor employs fewer and fewer directly, bringing difficulties of control, quality and training. A variety of management systems, including project management by intermediaries, a lack of acceptance of responsibility for control and the increasing complexity of design, materials and equipment have all contributed to the industry's continuing poor accident record.

For the most part, the hazards and technical solutions for hazard prevention and risk reduction are well-known. A possible exception is the general lack of awareness (and data) on the occupational health record of the industry's workers, who are potentially exposed to the elements as well as a wide variety of hazardous substances. The Construction Regulations are now over thirty years old, and were not designed to anticipate the technological developments over the period, or the considerable changes in management systems. Nevertheless, the statistics show that the injuries which happen today are largely the same as those of yesterday, happening for the same reasons.

The need for further change in management attitudes, employee attitudes, in regulations and codes is well-recognised and long overdue. Work by the Health and Safety Executive, parallelled by EC Directives to be in force shortly, should make significant inroads in the industry's health and safety problems, provided that they are backed by recognition of the need for change in companies of all sizes, by the self-employed and by enforcement. The degree of Trade Union involvement in the industry is less than it was, leading to calls from many involved for the election of safety representatives to be derestricted and open to all on site.

Meanwhile, the reader is referred to the Law Module for summaries of the present legal controls on many aspects of site work, and to the Safety Management Techniques Module for information on control techniques, which are as applicable to the construction industry as elsewhere. This Section will examine other aspects of the control of construction health and safety, after reviewing the available statistical evidence on causation. At the time of writing, proposals for activating the European Directive covering temporary and mobile construction sites are subject to consultation. Regulations to activate the Directive must be in place by January 1994, and will probably be known as the Construction (Design and Management) Regulations 1993. They will require greater liaison and planning between clients and those in the construction industry, and widen the definition of 'construction work' currently in use. The reader is strongly advised to become acquainted with CONDAM, especially as almost every employer will be a 'client' as defined, and thus a duty holder.

THE ACCIDENT RECORD

The construction industry's reported rates for fatalities and major injuries are the highest of the five main sectors of employment. They are five times higher than in manufacturing industry, and more than ten times higher than in service industries. In 1990/91, 12.2% of all reported injuries were reported in the construction sector, which at the time had an estimated 6.9% of the workforce. Just under 22,000 injuries were reported under the RIDDOR system.

Every day, on average, 59 construction workers are reported as injured. Studies suggest that under-reporting is very common, and that up to twice that number of injuries may occur in total. Under-reporting is thought to be particularly common amongst the self-employed, who reported 2,475 injuries in 1990/91 instead of the 30,000 or more which current conservative estimates believe to have occurred.

A better indication of the picture can be drawn from fatalities, which give a smaller statistical sample but are much less likely to be under-reported (except possibly those which are related to ill-health exposures).

Construction Safety

On average, a worker is killed in the industry every three days, and one member of the public is killed each month by construction activities. There are about 100 fatalities a year to workers, a figure which has remained fairly constant over recent years despite increased safety activity and a declining work population. However, in the early 1970s the rate was about twice its present level. The 1991 New Earning Survey indicated a decline of 2% in hours worked in the industry, and a slight decline in injury numbers is said by the Health and Safety Commission to reflect a reduction in activity levels.

Reported fatalities to the self-employed more than doubled in the four years between 1986/87 and 1989/90.

The **risk of injury** for someone working in the industry for twenty years is that have a 1 in 600 chance of being killed as a result of a work accident, and a 1 in 2 chance of suffering a reportable injury. The former risk compares with a 1 in 550 chance for agriculture, and a 1 in 2500 chance in manufacturing industry. These estimates are based on assumed numbers of injuries rather than those actually reported.

TYPES OF INJURY

The most common source of fatalities in the last five years has been the head injury, accounting for almost one third of the total. there is claimed to be a marked reduction, of about 25%, in the head injuries rate overall, following the introduction of the Construction (Head Protection) Regulations on 1st April 1990.

For major injuries reported, more than 75% were fractures, notably wrist, ankle, upper arm and lower leg. Over 3-day injuries were most often sprains and strains (36.9%) in 1989/90, just under half of which were back injuries.

ACCIDENT CAUSES

Firstly, Canadian studies have shown that active involvement in safety management by the most senior levels in a construction company is directly correlated with reductions in numbers of accidents and injuries.

Knowledge of causation patterns provides a starting point for focusing preventive measures. Case studies and descriptions of accidents can be used to give information about prevention techniques – the Health and Safety Executive's publication "Blackspot – Construction" is essential reading. It comments that in a sample studied, 90% of fatalities were found to be preventable, and in

70% of cases positive management action could have saved lives. Table 1 shows the distribution of causes in the sample. 74% of these fatalities happened in the 'building' sector, 26% in 'civil engineering'.

Whilst 561 employees were killed in the five-year period studied, a further 94 were killed who were self-employed and working as sub-contractors, 26 more were self-employed working on their own, 37 adult members of the public were killed, and 21 children. In total, 739 deaths were recorded, the highest in one year being 162, and the lowest annual total 137.

TABLE 1:

Fatalities to employees in the construction industry, 1981 – 1985 inclusive.

CATEGORIES OF ACCIDENT CAUSES	TOTAL	PERCENTAGE
Falls – Roof edge	39	6.95
Through fragile materials	63	11.23
From internal structure of roofs	5	0.89
From cradles, suspended chairs and baskets	11	1.96
During erecting and dismantling of scaffolding	22	3.92
During demolition work	24	4.28
From tower scaffolds	10	1.78
During steel erection	10	1.78
From scaffolds and working platforms	40	7.13
Ladders and stepladders	40	7.13
Other categories of falls	21	3.74
All falls	285	50.80
Insecure loads or equipment	32	5.70
Falls of rock or earth from sides of excavations and tunnels	35	6.24
Collapse of structures or parts of structures	45	8.02
Transport and mobile plant	115	20.50
Electrocution, including burns	21	3.74
Fires and explosions	7	1.25
Asphyxiation and drowning	18	3.21
Unclassified	3	0.53
TOTAL	561	(99.98)

Falls of all kinds account for about half of all fatalities in construction, although the category is so wide that an internal breakdown of the data is useful. The three worst **task areas** (75% of all deaths) were maintenance (42%), transport and mobile plant (20%) and demolition/dismantling (13%) – each receives detailed treatment in this book.

OCCUPATIONS MOST AT RISK

The same samples give information on the occupations of those killed. Whilst no data is given for the percentage distribution of occupations across the workforce, it can be seen that a) some are more at risk than others, b) some are at high risk relative to their assumed numbers (and may well be unaware of it). For example, there are over four times as many labourers killed as managers, but it can be assumed that there are considerably more than four times as many labourers as managers employed in the industry.

TABLE 2:

Occupations of those killed in construction work, 1981 – 1985 inclusive (including the self-employed)

OCCUPATION	TOTAL	PERCENTAGE
Labourers and civil engineering workers	213	31.30
Roofing workers	99	14.54
Painters	53	7.78
Drivers	51	7.49
Demolition workers	50	7.34
Managerial and professional	49	7.16
Carpenters/joiners	37	5.43
Scaffolders	23	3.38
Steel erectors	22	3.23
Bricklayers	19	2.79
Plumbers/glaziers	13	1.91
Electricians	11	1.62
Other	41	6.02
TOTAL	**681**	**(99.99)**

OCCUPATIONAL HEALTH AND HYGIENE

Traditionally the construction industry's high level of injury-causing accidents has received the attention of enforcement, media publicity and management action. Arguably, the size of that problem has led to a neglect of the less tangible consequences of occupational hygiene and health problems, apart from well-publicised topics such as asbestos. Reports suggest that construction workers age prematurely due to hypothermia caused by working in the cold and wet. Respiratory diseases such as bronchitis and asthma are also thought to occur at above average levels in construction workers.

The authors' experience with the implementation of the COSHH Regulations in construction shows that there is little awareness of the principles of assessment, or significant appreciation of the risks to workers from substances brought onto the site – and especially from those created there. Also, there is said to be a disappointing response from the industry to the noise controls (mostly managerial action and measurement requirements) imposed by the Noise at Work Regulations.

CONTROLLING CONSTRUCTION ACCIDENTS

The successful control of hazards and risks in the construction industry depends upon the same principles as in other industries. Specific organisational problems do exist, and need to be addressed, chiefly by recognising the need to appreciate health and safety matters at the very earliest stages – design, specification of materials, their delivery quantities and batch sizes, work planning, interaction between contractors, and between contractors and the public.

Legislative improvements, especially the coming Construction (Design and Management) Regulations (CONDAM), require specific control and planning to be done prior to and during the construction phase. Projects will require safety plans before work starts, on which tenders have been based. The plans, drawn up by a nominated Planning Supervisor, will be passed to the selected competent Principal Contractor, who has well-defined responsibilities for control of the safety of work on site, including the activities of other contractors. The safety plan will be revised by the Principal Contractor, in particular to incorporate the risk assessments provided by all contractors. It thus becomes a working document through the duration of the project. The safety file – another innovation – will be the statement of how the construction was done, for use of the client and future modifiers or demolishers of it.

The Management of Health and Safety at Work Regulations 1992 include requirements which have an impact upon construction work. The CONDAM Regulations will add further definition to the details for this industry, but the reader will find it instructive to review Module D Section 5 from the aspect of construction activity. Regulations 9 and 10 are of particular interest.

Method statements covering high-risk operations should be required in advance, from those doing the work. Each should contain acknowledgement that if operational conditions force any deviation from the method statement, this must be agreed by site supervi-

Construction Safety

sion and accepted by them in writing. Tasks where method statements should always be used include demolition, the use of explosives, erection of steel and structural frames, deep excavations and tunnelling, lifting with more than one appliance, use of suspended access equipment, and falsework. They should also be considered for roofwork, especially if fragile roofs are identified. Roofwork method statements should detail access methods, work procedure and fall protection systems. All method statements are plans for work which are based on risk assessments (See Module C Section 3).

Training in techniques and skills for workers should include a strong safety element. Management training is especially important as it fosters that positive commitment to safety which is necessary before the construction process can be successfully managed in safety.

Demolition

INTRODUCTION

Research studies show that accidents during demolition work are more likely to be fatal than those in other areas of construction work. The more significant causes of accidents which have high potential for serious injury are premature collapse of buildings and structures, and falls from working places and access routes. A common feature of demolition accidents is that investigation shows that there is usually a failure to plan the work sufficiently at the appropriate stage, which leaves operatives on site to devise their own methods of doing the work without knowledge and information about the dangers that confront them. Failure to plan is, of course, not confined to demolition or even the construction industry, but it is difficult to think of many other situations where the consequences are visited so rapidly upon the employees.

PLANNING FOR SAFETY

As much information should be obtained about the work to be done, at the earliest possible time. The extent to which a client is willing or able to provide structural information will partly depend upon whether it is actually available to him. In some cases the client may exercise a degree of control over the work and may thus be required to provide relevant information under the Health and Safety at Work etc. Act 1974 and the Management of Health and Safety at Work Regulations 1992. Information should be made available when inviting tenders to carry out the work – unless specifically agreed, as when it is not in the reasonable knowledge of the client, it should not be up to the contractor to discover a particular hazard of the structure or building to be demolished.

Demolition is normally carried out by specialist contractors with experience. Supervision and control of the actual work is required to be done by people experienced in demolition, and this may not be achieved by a general contractor. Clients (owners) should ensure that work of this type carried out on their behalf is done by contractors competent to recognise features of the work which are likely to cause complications or require further investigation at an early stage.

Clients (owners) should ensure that details and drawings are prepared and kept during the initial construction and cover use modifications, so that risks to health and safety can be assessed adequately and the appropriate precautions put in hand. Information on storage and use of chemicals can be obtained from the owner, but in cases where the site has been vacant for some time or previous ownership is unclear the services of a competent analyst may be needed.

An experienced structural engineer or surveyor should be employed, if the owner cannot provide structural details, building drawings or surveyors' reports, to obtain the necessary information. Architectural or archaeological items for retention must be defined before the preferred method of demolition is determined.

The client (owner) and the contractor(s) should agree the programme for demolition work and the preferred method, to ensure that necessary resources are available.

DEMOLITION SURVEYS

Using information supplied, prospective contractors should carry out a survey in sufficient detail to identify structural problems, and risks associated with flammable substances or substances hazardous to health. The precautions required to protect employees and members of the public from these risks, together with the preferred demolition procedure, should be set out in a **method statement** (see below).

The survey should:
- Take account of the whole site. Access should be permitted for the completion of surveys and information made available in order to plan the intended method of demolition.
- Identify adjoining properties which may be affected by the work – structurally, physically or chemically. Premises which may be sensitive to the work, other than domestic premises, include hospitals, telephone exchanges and industrial premises with machines vulnerable to dust, noise or vibration.
- Identify the need for any shoring work to adjacent properties or elements within the property to be demolished. Weatherproofing requirements for the work will also be noted.
- Identify the structural condition, as deterioration may impose restrictions on the demolition method.

The survey should preferably be divided into structural and chemical aspects, the latter noting any residual contamination. Structural aspects of the survey should note variations in the type of construction within individual buildings and amongst buildings forming a complex due for demolition. The person carrying out the survey should be competent. An assessment of the original construction method including any temporary works required can be very helpful.

Demolition

PREFERRED METHOD OF WORK

The basic ideal principle is that structures should be demolished in the reverse order to their erection. The method chosen should gradually reduce the height of the structure or building, or arrange its deliberate controlled collapse so that work can be completed at ground level.

Structural information obtained by the survey should be used to ensure that the preferred method of work retains the stability of the parts of the structure or building which have not yet been demolished. The aim should be to adopt methods which make it unnecessary for work to be done at height. If this cannot be achieved then systems which limit the danger of such exposure should be employed. The use of balling machines, heavy duty grabs or pusher arms may avoid the need to work at heights. If these methods are possible, the contractor must be satisfied that sufficient area is available for the safe use of the equipment, and that the equipment is adequate for the job.

When work cannot be carried out safely from part of a permanent structure or building, working platforms can be used. These include scaffolds, towers and power-operated mobile work platforms. Where these measures are not practicable, safety nets or harnesses properly anchored may be used.

Causing the structure or building to collapse by the use of wire ropes or explosives may reduce the need for working at heights, but suitable access and working platforms may be needed during the initial stages.

A knowledge of structural engineering principles is necessary to avoid premature collapse, especially an understanding of the effect of pre-weakening by the removal, cutting or partial cutting of structural members.

METHOD STATEMENTS

In order to comply with Section 2(2) of the Health and Safety at Work etc. Act 1974 (provision of a safe system of work), the production of a method statement is recognised as necessary for all demolition work. Because of the special demolition needs of each structure, an individual risk assessment must be made by the employer undertaking the work, in writing, to comply with Regulation 3 of the Management of Health and Safety at Work Regulations 1992. The detailed method statement will be derived from the risk assessment carried out by the contractor. This must be drawn up before work starts, and communicated to all

involved. It should identify the work procedure, associated problems and their solutions, and should form a reference for site supervision.

Method statements should be easy to understand, agreed by and known to all levels of management and supervision, including those of sub-contracting specialists.

The method statement should include:

1. The sequence of events and method of demolition (including drawings/diagrams) or dismantling of the structure or building

2. Details of personnel access, working platforms and machinery requirements

3. Specific details of any pre-weakening of structures which are to be pulled down or demolished using explosives

4. Arrangements for the protection of personnel and public, and the exclusion of unauthorised people from the work area. Details of areas outside the site boundaries which may need control during critical aspects of the work must be included

5. Details of the removal or isolation of electrical, gas and other services, including drains

6. Details of temporary services required

7. Arrangements for the disposal of waste

8. Necessary action required for environmental considerations (noise, dust, pollution of water, disposal of contaminated ground)

9. Details of controls covering substances hazardous to health and flammable substances

10. Arrangements for the control of site transport

11. Training requirements

12. Welfare arrangements appropriate to the work and conditions expected

13. Identification of people with special responsibilities for the co-ordination and control of safety arrangements.

DEMOLITION TECHNIQUES

Piecemeal demolition is done by hand, using hand-held tools, sometimes as a preliminary to other methods. Considerations include provision of a safe place of work and safe access/egress, and debris disposal. It can be completed or begun by machines such as balling machines, impact hammers or hydraulic pusher arms. Considerations for these include safe operation of the machines, clearances, capability of the equipment and protection of the operator.

Deliberate controlled collapse involves pre-weakening the structure or building as a preliminary, and completion by use of explosives or overturning with wire rope pulling. Considerations for the use of explosives include competence, storage, blast protection, firing programmes and misfire drill. Wire rope pulling requires a similar level of expertise, as well as selection of materials and clear areas for rope runs.

DEMOLITION TRAINING

The importance of adequate training in demolition has now been recognised by the introduction of the Construction Industry Training Board's "Scheme for the Certification of Competence of Demolition Operatives". Training requirements are imposed by the Construction (General Provisions) Regulations 1961, the Health and Safety at Work etc. Act 1974, and specifically by the Management of Health and Safety at Work Regulations 1992.

LEGAL REQUIREMENTS

Several Acts and Regulations control demolition work. At the time of writing, the demolition of a building is defined as a "building operation", and the demolition of most other structures is classed as "works of engineering construction" in Section 176 of the Factories Act 1961. Section 127 applies certain provisions of the Act to demolition – the need to notify work, keep a General Register, post abstracts of Acts etc. The Health and Safety at Work etc. Act 1974 applies whenever demolition work is done, as do the Management of Health and Safety at Work Regulations 1992 and others introduced on 1st January 1993.

The five sets of Construction Regulations apply to all demolition work – the Sections in this book dealing with those Regulations should be consulted for a summary. Other Acts and Regulations which may be applicable include those covering COSHH, Electricity, First-Aid, Asbestos, Lead, Woodworking Machines, Abrasive Wheels, Control of Pollution and Environmental Protection.

REVISION

Six elements in planning for safety in demolition:

- Information
- Demolition surveys
- Preferred method of work
- Method statements
- Consultation
- Training

Two main demolition techniques:

- Piecemeal – by hand or machine
- Deliberate controlled collapse, including pre-weakening, using explosives or overturning

Selected References

FOR ALL SECTIONS

Health and Safety at Work etc Act 1974
Management of Health and Safety at Work Regulations 1992
Workplace (Health, Safety and Welfare) Regulations 1992
Provision and Use of Work Equipment Regulations 1992
Manual Handling Operations Regulations 1992
Personal Protective Equipment at Work Regulations 1992
Factories Act 1961
Offices, Shops and Railway Premises Act 1963
Fire Precautions Act 1971
and associated Approved Codes of Practice and Guidance on Regulations

SAFE MACHINERY DESIGN AND GUARDING

Abrasive Wheels Regulations 1970
Construction (General Provisions) Regulations 1961
Horizontal Milling Machines Regulations 1928
Mines and Quarries Act 1954
Power Presses Regulations 1965
Woodworking Machines Regulations 1974

HSE publications

L22	Work equipment – guidance on Regulations
PM1	Guarding of portable pipe threading machines
PM2	Guards for planing machines

British Standard

BS5304	Code of Practice for safety of machinery

MECHANICAL HANDLING

Construction (Lifting Operations) Regulations 1961

British Standards

BS7121	Code of Practice for the safe use of cranes
BS5667	Specification for continuous mechanical handling equipment – safety requirements
BS4531	Specification for portable and mobile traughed belt conveyors
BS5667	Part 18 – Conveyors and elevators with chain elements – guarding
	Part 19 – Belt conveyors – guarding

Guidance (HSC/E)

PM3	Erection and dismantling of tower cranes
PM9	Access to tower cranes
PM55	Safe working with overhead travelling cranes
HS(G)6	Safety in work with lift trucks

MANUAL HANDLING

HSE Publications

L23	Manual handling – guidance on Regulations
HS(G)48	Human factors in industrial safety
HS(G)60	Work-related upper limb disorders – a guide to prevention
	Watch your step – prevention of slipping, tripping and falling accidents at work

ACCESS EQUIPMENT

Construction (Working Places) Regulations 1966
Agricultural (Ladders) Regulations 1957

British Standards

BS1129	Timber ladders, steps, trestles and lightweight stagings
BS1139	Metal scaffolding
BS2037	Aluminium ladders, steps and trestles
BS2482	Timber scaffold boards
BS2830	Suspended safety chairs and cradles
BS5323	Code of Practice for scissor lifts
BS5973	Code of Practice for access and working scaffolds and special structures in steel
BS5974	Code of Practice for temporarily installed suspended scaffolds and access equipment
BS6289	Part 1 – Code of Practice for mobile scissor operated work equipment
BS6307	Permanently installed suspended access equipment
BS7171	Specification for mobile elevating work platforms

Guidance (HSC/E)

GS15	General access scaffolds
GS31	Safe use of ladders, stepladders and trestles
GS42	Tower scaffolds
PM30	Suspended access equipment
HS(G)19	Safety in working with power operated mobile work platforms
HS(G)23	Safety at power operated mast work platforms

TRANSPORT SAFETY

Dangerous Substances (Conveyance by Road in Road Tankers and Tank Containers) Regulations 1981

Road Traffic (Carriage of Dangerous substances in Packages etc.) Regulations 1986

British Standards

BS5073   Guide to storage of goods in freight containers

BS6939   Recommendations for intermediate bulk containers (IBC's) for dangerous goods

Guidance (HSC/E)

COP 11   Operational provisions of the Dangerous Substances (Conveyance by Road in Road Tankers and Tank Containers) Regulations 1981

COP 14   Road tanker testing

COP 17   Operational provisions of the Road Traffic (Carriage of Dangerous Substances in Packages etc.) Regulations 1986

HS(G)26   Transport of dangerous substances in tank containers

HS(R)13   A guide to the Dangerous Substances (Conveyance by Road in Road Tankers and Tank Containers) Regulations 1981

HS(R)24   A guide to the Road Traffic (Carriage of Dangerous Substances in Packages etc.) Regulations 1986

CLASSIFICATION OF DANGEROUS SUBSTANCES

Classification, Packaging and Labelling of Dangerous Substances Regulations 1984 (amended 1989)

Notification of New Substances Regulations 1982 (amended 1986)

Guidance (HSC/E)

COP 22   Classification and labelling of substances dangerous for supply: Notification of New Substances Regulations 1982: Classification, Packaging and Labelling of Dangerous Substances Regulations 1984

Information approved for the classification, packaging and labelling of dangerous substances for supply and conveyance by road (Third Edition)

HS(G)27   Substances for use at work: the supply of information

HS(R)1   Packaging and labelling of dangerous substances regulations and guidance notes

HS(R)14   A guide to the Notification of New Substances Regulations 1982

HS(R)22   A guide to the Classification, Packaging and Labelling of Dangerous Substances Regulations 1984

CHEMICAL SAFETY

Control of Substances Hazardous to Health Regulations 1988

Control of Industrial Major Accident Hazards Regulations 1984

Dangerous Substances (Notification and Marking of Sites) Regulations 1990

Notification of Installations Handling Hazardous Substances Regulations 1982

Guidance (HSC/E)

COP 29   COSHH and control of carcinogenic substances (see also COP's 30,31 and 41)

HS(G)25   CIMAH 1984 – further guidance on emergency plans

HS(R)16   A guide to the Notification of Installations Handling Hazardous Substances Regulations 1982

HS(R)21   A guide to the Control of Industrial Major Accident Hazards Regulations 1984

HS(R)29   A guide to the Dangerous Substances (Notification and Marking of Sites) Regulations 1990

ELECTRICITY AND ELECTRICAL EQUIPMENT

Electricity at Work Regulations 1989

Electrical Equipment for Explosive Atmospheres (Certification) Regulations 1990

Low Voltage Electrical Equipment (Safety) Regulations 1989

British Standards

BS2754   Memorandum: Construction of electrical equipment for protection against electric shock

BS4343   Industrial plugs, socket-outlets and couplers for AC and DC supply

Selected References

Guidance (HSC/E)

HS(R)25	Memorandum of Guidance on the Electricity at Work Regulations
HS(G)13	Safety in electrical testing
HS(G)22	Electrical apparatus for use in potentially explosive atmospheres
HS(G)47	Avoiding danger from underground services
GS6	Avoidance of danger from overhead electric lines
GS24	Electricity on construction sites
GS27	Protection against electric shock
GS37	Flexible leads, plugs and sockets etc.
PM32	Safe use of portable electrical apparatus (electrical safety)
PM38	Selection and use of electric handlamps
PM64	Electrical safety in arc welding

FIRE

Fire Certificates (Special Premises) Regulations 1976
Fire Precautions (Application for a Certificate) Regulations 1989
Fire Precautions (Non-certificated Factory, Office, Shop and Railway Premises)(Revocations) Regulations 1989
Fire Safety and Safety of Places of Sport Act 1987

British Standard

BS5423   Specification for portable fire extinguishers

Guidance (Home Office)

Code of Practice: Fire Precautions in Factories, Offices, Shops and Railway Premises not required to have a fire certificate

Fire Precautions Act 1971: Guide to fire precautions in existing places of work that require a fire certificate

Guide to fire precautions in existing places of entertainment and like places

CONSTRUCTION SAFETY

Construction Regulations 1961, 1966 and 1989

Guidance (HSE)

Managing health and safety in construction; principles and application to main contractor/subcontractor projects – Construction Industry Advisory Committee

Blackspot Construction – a study of 5 years' fatal accidents in the building and civil engineering industries

Construction summary sheets SS1 – 18

GS28/1	Safe erection of structures – Part 1 Initial planning and design
GS28/2	Safe erection of structures – Part 2 Site management and procedures
GS28/3	Safe erection of structures – Part 3 Working places and access
GS28/4	Safe erection of structures – Part 4 Legislation and training
HS(G)33	Safety in roofwork

Construction Safety – manual produced by Building Employers Confederation

DEMOLITION

Construction Regulations 1961, 1966 and 1989

Guidance (HSE):

GS29/1	Health and safety in demolition work Part 1 – preparation & planning
GS29/2	Health and safety in demolition work Part 2 – legislation
GS29/3	Health and safety in demolition work Part 3 – techniques
GS29/4	Health and safety in demolition work Part 4 – health hazards

British Standard

BS 6187   Code of Practice – demolition

Module B:
Occupational Health
and Hygiene

Introduction to Occupational Health and Hygiene

Module B Section I

Occupational health anticipates and prevents health problems which are caused by the work which people do. In some circumstances the work may aggravate a pre-existing medical condition, and stopping this is also the role of occupational health. Health hazards often reveal their effects on the body only after the passage of time; many have cumulative effects, and in some cases the way this happens is still not fully understood. Because the effects are often not immediately apparent, it can be difficult to understand and persuade others that there is a need for caution and control. Good occupational hygiene practice encompasses the following ideas:

- **Recognition** of the hazards or potential hazards
- **Quantification** of the extent of the hazard – usually by measuring physical/chemical factors and their duration, and relating them to known or required standards
- **Assessment** of risk in the actual conditions of use, storage, transport and disposal
- **Control** of exposure to the hazard, through design, engineering, working systems, the use of personal protective equipment and biological monitoring
- **Monitoring change** in the hazard by means of audits or other measurement techniques, including periodic re-evaluation of work conditions and systems.

HISTORICAL DEVELOPMENT

There is evidence that the Greeks and Romans were aware of the hazards and risks to health, not to mention safety, in work activity, especially in the mines and extraction processes. Major milestones in the history of occupational health up to the beginning of the present century are as follows:

1526 Georg Bauer (in Latin texts Georgius Agricola) appointed as physician to the miners of Joachimsthal, and recommended mine ventilation, also the use of veils over faces to protect the miners from harmful dust. His treatise on mining of metals, including health aspects, **De Re Metallica**, was published in 1556.

1567 Von Hohenheim (in Latin texts Paracelsus) had a monograph published after his death on the lung diseases of miners and smelters.

1700 The Italian physician Bernardino Ramazzini wrote the first comprehensive document on occupational health, **De Morbis Artificium Diatriba**, a history of occupational diseases. Working in Padua, he was the first to suggest that physicians should ask their patients about their work when diagnosing illness.

1802 The Health and Morals of Apprentices Act passed in Great Britain, the world's first occupational health and safety legislation.

1831 Dr. Charles Thackrah, early UK occupational health pioneer, published **"The Effects of Principal Arts, Trades and Professions... on Health and Longevity"**.

1898 Thomas Legge (knighted for his work in 1925) appointed as first Medical Inspector of Factories.

HEALTH HAZARDS

Health hazards can be divided into four broad categories: physical, chemical, biological and ergonomic. Examples of the categories are:

- **Physical** – Air-pressure, heat, dampness, noise, radiant energy, electric shock.
- **Chemical** – Exposure to toxic materials such as dusts, fumes and gases
- **Biological** – Infection, e.g. tetanus, hepatitis and legionnaire's disease
- **Ergonomic** – Work conditions, stress, man-machine interaction.

TOXICITY OF SUBSTANCES

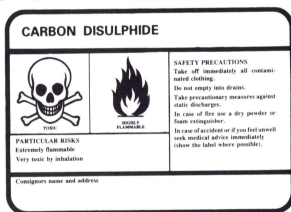

Toxicity is the ability of a substance to produce injury once it reaches a site in or on the body. The degree of harmful effect which a substance can have depends not only upon its inherent harmful properties but also upon the **route** and the **speed** of entry into the body. Substances may cause health hazards from a single exposure, even for a short time (**acute effect**) or after prolonged or repeated exposure (**chronic effect**). The substance may affect the body at the point of contact, when it is known as a **local agent**, or at some other point, when it is described as a **systemic agent**.

Introduction to Occupational Health and Hygiene

Absorption is said to occur only when a material has gained access to the blood stream and may consequently be carried to all parts of the body.

WHAT MAKES SUBSTANCES TOXIC?

The effect a substance will have on the body cannot always be predicted with accuracy, or explained solely on the basis of physical and chemical laws. The influence of the following factors combine to produce the **effective dose** (see Section 4):

Quantity or **concentration** of the substance
The **duration** of exposure
The **physical state** of the material, e.g. particle size
Its **affinity** for human tissue

Its **solubility** in human tissue fluids
The **sensitivity** to attack of human tissue or organs

LONG TERM AND SHORT TERM EXPOSURE

Substances which are toxic can have a toxic effect on the body after only one single, short exposure. In other circumstances, repeated exposure to small concentrations may give rise to an effect. A toxic effect related to an immediate response after a single exposure is called an **acute effect**. Effects which result after prolonged (hours or days or much longer) are known as **chronic effects**. "Chronic" implies repeated doses or exposures at low levels; they generally have delayed effects and are often due to unrecognised conditions which are therefore permitted to persist.

SELF-ASSESSMENT QUESTIONS

1. List the possible sources of health hazards in your workplace.

2. Operatives in your workplace report general discomfort when working with a material. Discuss how you would assess the problem.

REVISION

Four main health hazards:

- Physical
- Chemical
- Biological
- Ergonomic

Six factors determine toxicity:

- Concentration
- Duration
- Physical state
- Affinity
- Solubility
- Sensitivity

Acute effects are immediate responses to single short-term exposures

Chronic effects are long term responses to prolonged exposures

INTRODUCTION

The body's response against the invasion of substances likely to cause damage can be divided into external or **superficial** defences and internal or **cellular** defences. These defence mechanisms inter-relate, in the sense that the defence is conducted on a number of levels at once, and not in a stage-by-stage pattern.

SUPERFICIAL MECHANISMS OF DEFENCE

The superficial mechanisms work by the action of cell structures, such as organs and functioning systems.

The body's largest organ, the **skin**, provides a useful barrier against many foreign organisms and chemicals (but not against all of them). Its effect is, of course, limited by its physical characteristics. Openings in the skin, including sweat pores, hair follicles and cuts can allow entry, and the skin itself may be permeable to some chemicals such as toluene. The skin can withstand limited physical damage because of its elasticity and toughness, but its adaptation to cope with modern substances is usually viewed by its owner as unhelpful – dermatitis, with thickening and inflammation, is painful and prominent.

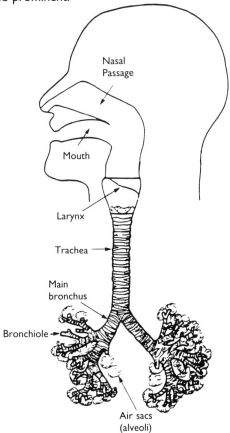

Defences against inhalation of substances harmful to the body begin in the **respiratory tract**, where a series of reflexes activate the coughing and sneezing mechanisms to expel forcibly the triggering substance. Many substances and micro-organisms are successfully trapped by nasal hairs and the mucus lining the passages of the respiratory system. The passages are also well supplied with fine hairs which sweep rythmically towards the outside and pass along larger particles. These hairs form the **ciliary escalator**. The respiratory system narrows as it enters the lungs, where the ciliary escalator assumes more and more importance as the effective defence. In the deep lung areas, only small particles are able to enter the alveoli (where gas exchange with the red blood cells takes place), and cellular defence predominates there.

For **ingestion** of substances entering the **mouth and gastrointestinal tract**, saliva in the mouth and acid in the stomach provide useful defences to substances which are not excessively acid or alkaline, or in great quantity. The wall of the gut presents an effective barrier against many insoluble materials. Vomiting and diarrhoea are additional reflex mechanisms which act to remove substances or quantities which the body is not equipped to deal with without damage to itself. Thus, there are a number of primitive defences, useful at an earlier evolutionary stage to prevent man unwittingly damaging himself, which are now available to protect against a newer range of problems as well as the old.

Eyes and ears are potential entry routes for substances and micro-organisms. The eyes prevent entry of harmful material by way of the eyelids, eye lashes, conjunctiva (thin specialised outer skin coating of the eyeball), and by bacteria-destroying tears. The ears are protected by the outer shell or pinna, and the ear drum is a physical barrier at the entrance to the sensitive mechanical parts and the organ of hearing. Waxy secretions protect the ear drum, and trap larger particles.

Other orifices may be invaded by micro-organisms. Generally acid environments, such as in the urethra, do not promote growth. Sexual contact is the main source of exposure.

CELLULAR MECHANISMS OF DEFENCE

The cells of the body possess their own defence systems.

Prevention of excessive blood loss from the circulation through blood clotting and coagulation

Body Response

prevents excessive bleeding and slows or prevents the entry of germs into the blood system.

Phagocytosis is the scavenging action of a defensive body cell (white blood cell) against an invading particle. A variety of actions can be used, including chemical, ingestion, enzyme attack and absorption.

Secretion of defensive substances is done by some specialised cells. Histamine release and heparin, which promotes availability of blood sugar, are examples.

Inflammatory response can isolate infected areas, remove harmful substances by an increased blood flow to the area, and promote the repair of damaged tissue.

Repair of damaged tissue is a necessary defence mechanism, which includes the removal of dead cells, increased availability of defender cells, and replacement of tissue strength and soundness by means of temporary and permanent repairs e.g. scar tissue.

Immune response is the ability to resist almost all organisms or toxins that tend to damage tissues. Some immunity is **innate**, such as the phagocytosis of organisms and their destruction by acid in the gut. In addition, the human body has the ability to develop extremely powerful specific immunity against invading agents. This is **acquired** immunity, also known as **adapting** immunity. Acquired immunity is highly specific, the resistance developing days or weeks after exposure to the invading agent.

SELF-ASSESSMENT QUESTIONS

1. Describe how the body can defend and repair itself when the skin is cut.

2. Give examples of reflexes which take part in body response to the presence of foreign substances.

REVISION

The body's response to potentially harmful substances and micro-organisms can be:

SUPERFICIAL   OR   CELLULAR

- Respiratory tract
- Mouth & gut
- Skin
- Eyes & ears
- Other orifices

- Prevention of blood loss
- Phagocytosis
- Secretion of defensive substances
- Inflammatory response
- Repair of damaged tissue
- Immune response

ROUTES

Substances harmful to the body may enter it by three main routes. These are:

Absorption – through the skin, including entry through cuts and abrasions, and the conjunctiva of the eye. Organic solvents are able to penetrate the skin, as a result of accidental exposure to them or by washing. Tetraethyl lead and toluene are examples.

Ingestion – through the mouth, which is generally considered to be a rare method of contracting industrial disease. However, the action of the main defence mechanisms protecting the lungs rejects particles and pushes them towards the mouth, and an estimated 50% of the particles deposited in the upper respiratory tract and 12.5% from the lower passages are eventually swallowed.

Inhalation – the most important route of entry, which can allow direct attacks against lung tissue which bypass other defences such as those of the liver. The lungs are very efficient in transferring substances into the body from the outside environment, and this is way inside for 90% of industrial poisons.

RESULTS OF ENTRY

Having gained entry into the body, substances can have the following effects:

Cause diseases of the skin such as:

Non-infective dermatitis – an inflammation of the skin especially on hands, wrists and forearms. This can be prevented by health screening, good personal hygiene, use of barrier creams and/or protective clothing.

Scrotal cancer – produced by rubbing contact with workers' clothing impregnated with a carcinogen such as mineral oil, in close contact with the scrotum. This can be prevented by substitution of the original substance, by use of splash guards, and by provision of clean clothing and washing facilities for soiled work clothing.

Cause diseases of the respiratory system such as:

Pneumoconioses – resulting from exposure to dust which deposits on the lung, such as metal dust and man-made mineral fibre. Other examples of these fibroses of the lungs are **silicosis** due to the inhalation of free silica, and **asbestosis** from exposure to asbestos fibres.

Humidifier fever – giving influenza-like symptoms and resulting from contaminated humidifying systems.

Legionnaire's disease – from exposure to legionella bacteria.

Cause cancer and birth defects – by encouraging cells to undergo fundamental changes by altering the genetic material within the cell. Substances which can do this are **carcinogens**, which cause or promote the development of unwanted cells as cancer. Examples are asbestos, mineral oil, hard wood dusts and arsenic. **Teratogens** cause birth defects by altering genetic material in cells in the reproductive organs, and cause abnormal development of the embryo. Examples are organic mercury and lead compounds. **Mutagens** trigger changes affecting future generations.

Cause asphyxiation – by excluding oxygen or by direct toxic action. Carbon monoxide does this by competing successfully with oxygen for transport in the red cells in the blood.

Cause central nervous system disorders – by acting on brain tissue or other organs, as in the case of alcohol eventually causing blindness.

Cause damage to specific organs – such as kidneys and liver. An example is vinyl chloride monomer (VCM).

Cause blood poisoning – and producing abnormalities in the blood, as in benzene poisoning, where anaemia or leukaemia is the result.

SELF-ASSESSMENT QUESTIONS

1. "Ingestion of toxic chemicals is a rare method of contracting industrial disease". Discuss this statement, giving examples from your own experience.

2. Outline the measures which can be taken to prevent outbreaks of dermatitis.

REVISION

Three main routes of entry:

- Absorption
- Ingestion
- Inhalation

Occupational Exposure Limits

INTRODUCTION

An important part of an occupational hygiene programme is the measurement of the extent of the hazard. This is generally done by measuring physical and/or chemical factors, including exposure duration, and relating them to occupational hygiene standards.

Authorities in several countries publish recommended standards for airborne gases, vapours, dusts, fibres and fumes. The two primary (English language) sources are:

> **The Health and Safety Executive (HSE)** in the United Kingdom, which publishes Occupational Exposure Limits annually and as necessary. The UK Standard is essentially in two parts, specifying Maximum Exposure Limits (MELs) and Occupational Exposure Standards (OESs).

> **The American Conference of Governmental Industrial Hygienists (ACGIH)**, which publishes a list of Threshold Limit Values (TLVs) annually. The **Occupational Health and Safety Administration (OSHA)** publishes national Standards based on recommendations from the National Institute of Occupational Safety and Health (NIOSH).

MELs AND OESs – FOR THE UK

Maximum Exposure Limit (MEL) is the maximum concentration of an airborne substance, averaged over a reference period (e.g. 8-hour long-term) to which employees may be exposed by inhalation under any circumstances. Some substances have been assigned short-term MELs (e.g. 10-minute reference period). These substances give rise to acute effects and therefore these limits should never be exceeded.

Occupational Exposure Standard (OES) is the concentration of an airborne substance, averaged over a reference period, at which there is no current evidence that repeated (day after day) exposure by inhalation will be injurious to the health of employees. OESs should not be exceeded, but where this occurs effective steps should be taken as soon as practicable to reduce the exposure.

These values are given in units of parts per million (ppm) and milligrams per cubic metre (mgm^{-3}). They are given for two periods; long-term exposure (8-hour time-weighted average (TWA)) and short-term exposure (10-minute TWA). Some substances are designated "SK" which indicates that they can be absorbed through the skin.

GUIDANCE NOTE EH40

Health and Safety Executive Guidance Note EH40 is published annually, reproducing the current statutory list of Maximum Exposure Levels (MELs) and approved Occupational Exposure Standards (OESs). It is obviously important to see that the current document is the one consulted. This is identified by a suffix to the number EH40, showing the year of currency.

The exposure of employees to substances hazardous to health is to be prevented or, where this is not reasonably practicable, adequately controlled as a fundamental requirement of the COSHH Regulations (See Law Module D, Section 14). EH40 lists the limits to be used in determining the adequacy of the control of exposure by inhalation.

Advice and data in EH40 is to be taken in the context of the COSHH Regulations, especially Regulation 6 (Assessment), Regulation 7 (Control of exposure), Regulations 8 and 9 (Use and maintenance of control measures) and Regulation 10 (Monitoring of exposure. There is separate legislation for lead and asbestos, and these are not covered in detail in EH40. Exposure to micro-organisms and that below ground in mines is not dealt with.

USE OF EH40

Regulation 7 of the COSHH Regulations sets out requirements for the use of MELs and OESs for the purpose of achieving adequate control. Regulation 7(4) requires that where there is exposure to a substance for which an MEL has been assigned, the control shall, so far as inhalation of that substance is concerned, only be treated as adequate if the level of exposure is reduced as far as is reasonably practicable and in any case below the MEL.

Regulation 7(5) of the COSHH Regulations requires that, without prejudice to the generality of Regulation 7(1) where there is exposure to a substance for which an OES has been approved, the control of exposure shall, so far as the inhalation of that substance is concerned, be treated as adequate if:

- the OES is not exceeded, or
- where the OES is exceeded, the employer identifies the reasons for the standard being exceeded and takes appropriate action to remedy the situation as soon as is reasonably practicable.

LAYOUT OF EH40

The layout of the document has been standardised. It contains four sets of tables:

Table 1 List of MELs
Table 2 List of approved OESs
Table 3 Proposed changes to the list of approved OESs
Table 4 Substances to be reviewed

THRESHOLD LIMIT VALUES

The **Threshold Limit Value** is the concentration of an airborne substance and represents conditions under which it is believed that nearly all workers may be repeatedly exposed day after day without adverse health effects. There are different types of TLV:

Time Weighted Average TLV (TLV-TWA) – limits for indefinitely continued exposure 8 hours a day, 5 days a week.

Short Term Exposure Level TLV (TLV-STEL) – maximum concentrations of contaminant in air, beyond which the worker should not be exposed for more than a continuous exposure time period of 15 minutes.

Ceiling TLV (TLV-C) – this converts the TLV-TWA into a value not to be exceeded at any time.

All values quoted are for inhalation, and the units are milligrams per cubic metre (mgm^{-3}) or parts per million (ppm). Skin absorption is also denoted with "SK". Threshold Limit Values were discontinued in Great Britain because of disagreement with ACGIH over some values, and the requirement of ACGIH that the table of TLV values must be published as a whole or not at all. For practical purposes, there is no difference between TLVs and OESs.

It should be remembered that all occupational exposure limits refer to healthy adults working at normal rates over normal shift durations and patterns. In practice, it is advisable to work well below the standards set, and to bear in mind the philosophy of progressive risk reduction over time which is indicated by the newest Regulations introduced in response to European Directives.

SELF-ASSESSMENT QUESTIONS

1. What are the sources of occupational exposure limits?

2. Choose two substances found in your workplace air as contaminants, and look up their OESs or TLVs.

REVISION

UK standards:

- OESs
- MELs

USA values:

- TLV-TWA
- TLV-STEL
- TLV-C

Environmental Monitoring

INTRODUCTION

The key to preventing exposure to substances which could be hazardous to health depends upon the first two steps mentioned in Section 1 – recognition of the hazard or potential hazard, and evaluation of the extent of the hazard. People in the workplace may encounter hazards from several sources. An important means of evaluation is measurement to determine the extent of the threat.

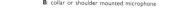

Microphone supported from hat or helmet

B: Microphone clipped to collar

A: about 4-10cm from side of head

A: head mounted microphone
B: collar or shoulder mounted microphone

For analysis of whether an area is safe to enter – direct reading instruments, except for particle qualitative and quantitative analysis.

MEASUREMENT – WHICH TECHNIQUE?

As discussed in Section 1, the health effects of exposure to toxic substances can be acute or chronic. It will therefore be necessary to distinguish appropriate types of measurement:

1. Long term measurements which assess the average exposure of a person over a given time period.

2. Continuous measurements capable of detecting short term exposure to high concentrations of contaminants which cause an acute exposure.

3. Spot readings can be used to measure acute hazards if the exact point of time of exposure is known and the measurement is taken at that time. Chronic hazards may be assessed if a statistically significant number of measurements are made.

SOME USEFUL DEFINITIONS

Dusts are solid particles suspended in air, which will settle under gravity. They are generated usually by mechanical handling processes including crushing and grinding. They can be of organic or inorganic origin. Particle size lies between 0.5 and 10 microns.

Fumes are solid particles formed by condensation from the gaseous state, for example metal oxides from volatilised metals. They can flocculate and coalesce. Their particle size is between 0.1 and 1 micron.

Mists are suspended liquid droplets formed by condensation from the gaseous state or by breakup of liquids in air. They can be formed by splashing, foaming or atomising – their particle size lies between 5 and 100 microns.

Fogs are fine mists comprised of suspended liquid droplets at the lower end of the particle size range.

Vapours are the gaseous forms of substances which are normally in the solid or liquid state, for example sodium vapour in luminaires. They are generated by decrease from normal pressure or temperature increase.

Gases are any substances in the physical condition of having no definite volume or shape, but tending to expand to fill any container into which they are introduced.

Aerosol is a term used to describe airborne particles which are small enough to float in air. These can be liquids or solids.

Smoke contains incomplete combustion products of organic origin, the particle sizes of which range between 0.01 and 0.3 microns.

NB. A **micron** is a unit of length corresponding to one millionth of a metre.

SUMMARY OF MEASUREMENT TECHNIQUES

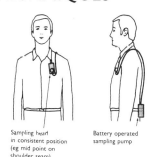

Sampling head in consistent position (eg mid point on shoulder seam)

Battery operated sampling pump

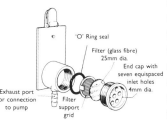

Exhaust port for connection to pump

'O' Ring seal

Filter (glass fibre) 25mm dia.

End cap with seven equispaced inlet holes 4mm dia.

Filter support grid

For chronic hazards – continuous personal dose measurement, continuous measurement of average background levels, spot readings of contaminant levels at selected positions and times.

For acute hazards – continuous personal monitoring with rapid response, continuous background monitoring with rapid response, spot readings of background contaminant levels at selected positions and times.

MEASUREMENT TECHNIQUES

The more common air quality measuring techniques are:

Grab sampling

Stain detectors are used for measuring airborne concentrations of gases and vapours. Well-known as a means of assessing alcohol consumption in roadside police checks, stain tubes are sealed glass tubes packed with chemicals which react specifically with the air contaminant being measured. In use, the tube is opened at the ends, a hand pump is attached and a standard volume of contaminated air is drawn through the tube. The chemical in the tube then changes colour in the direction of the airflow. The tube is calibrated so that the colour change corresponds to the concentration of the contamination.

The **main drawbacks** of grab sampling include inability to measure personal exposure by this method, except in the most general sense, and tube errors. These may arise because of the small volumes used, conditions such as temperature which affect some reactions in the tube chemical, and the possibility that the extent of the reaction may be influenced by the presence of other substances in the sampled air which also cause a colour change in the tube. The accuracy of this method is not high, and it is best to use it to give a rough indication of the presence of a contaminant, with an estimate of the extent of contamination. It does not provide a time-weighted average result, and a single reading may indicate a longer-term concentration.

Long Term Sampling

This involves the taking of air samples for several hours, thus giving the average concentration at which the contaminant is present throughout the sampling period. Sampling may be done by attaching equipment to the operator so as to sample air entering the breathing zone **(personal sampling)**, or by measuring at different points in the workplace (**static** or **area** sampling). Long term sampling can be done by the use of long term stain detector tubes which are connected to a pump. This draws air through the tube at a predetermined constant rate. At the end of the sampling period, the tube is examined and produces a value for the average level of concentration during the period.

The **main drawbacks** are as before, except that the accuracy of measurement of the sampled volume of air is improved by the use of a pump taking small samples over a long period.

For a few substances, direct measuring diffusion tubes are available, which do not require the use of a pump as diffusion is the means by which the sample of contaminant is collected.

More accurate results are obtained from operator personal sampling by means of **charcoal tube sampling**. This involves drawing air through a tube containing activated charcoal which absorbs the contaminant. The tube is then analysed in a laboratory to find the airborne concentration of the contaminant. Diffusion badges or monitors containing absorbents are becoming widely used. No pump is required; the method is reliable, versatile and accurate. Analysis is normally done in a specialist laboratory, so personal sampling generally suffers from delay in obtaining the results as compared with the use of stain detector tubes.

Dust sampling

The most widely used technique is simple dust filtration, involving the use of a small pump to suck a measured quantity of air through a filtering membrane over a period of time. The sampling head containing the membrane is then removed from the pump and analysed in a laboratory.

Direct monitoring

Some instruments available commercially, and also able to be hired, can produce an immediate quantitative analysis of the level of a particular contaminant, or even a qualitative analysis of the air sample as a whole. The results are available on a meter or chart recorder. Infra-red gas analysers are the most common type. This method allows the detection of the presence of short-term peak concentrations of contaminant during the work period, which is useful for working out control methods.

Hygrometers

These are instruments used for the measurement of water vapour in air. Although everyone is aware when the atmosphere becomes humid, peoples' ability to estimate humidity accurately is not good. Hygrometers can be very useful in measuring humidity and comfort of the working environment.

Measurement of other environmental hazards which may be encountered involves similar considerations and techniques, although the equipment required is usually more sophisticated. **Radiation** is measured by grab sampling using a Geiger counter; **microwave** energy can be measured by a meter as it is radiated, or its

Environmental Monitoring

presence can be detected by the fluorescing of a rare earth in a vacuum tube – an ordinary fluorescent lighting tube can be used for the purpose if small enough. **Sound** energy is measured using a proprietary meter, as is **light** for quantity and colour temperature.

Measurement of wave and particle energy, including noise, normally requires special equipment, and special training to operate it correctly and to produce reliable and useful results. Simple equipment can be very useful in identifying the presence of a hazard, but should not be relied upon totally in the development of controls.

An example is in the selection of personal hearing protection (see Section 8), where although a simple meter can identify the broad extent of the hazard, an octave band analysis would be required in order to match the characteristics of the sound source with the attenuating capabilities of different hearing protection. Note that this matching of the protective equipment to the risk it will control and to the wearer is a requirement of the Personal Protective Equipment at Work Regulations 1992 (See Module D, Section 10).

INTERPRETING THE RESULTS

Interpretation of the results is a skilled task, and involves making judgments about the results and the norms and standards laid down. The interpretation will determine the control strategy (see Section 4).

SELF-ASSESSMENT QUESTIONS

1. Explain the advantages of monitoring air quality standards using stain detector tubes.

2. Previously, you found exposure limit values for two substances present in your workplace air. Find out what methods are, or can be, used to measure how much of each is present?

REVISION

The main forms of environmental monitoring in the workplace are:

- Grab sampling
- Long term sampling
- Direct monitoring

Environmental Engineering Controls

INTRODUCTION

Having established a potential for injury in the workplace, selection of one or more control measures is necessary. An integral part of the effectiveness of the control is the monitoring of the controls once in place. It is important to remember that engineering and designing the problems out must be the primary consideration. The use of personal protective equipment is low in the list of control measures to be considered as a single solution. The following control measures can be used, in descending order of efficacy and priority:

- Substitution
- Isolation
- Enclosure of the process
- Local exhaust ventilation
- General (dilution) ventilation
- Good housekeeping
- Reduced exposure time
- Training
- Personal protective clothing and equipment
- Welfare facilities
- Medical surveillance

The use of warnings, such as signs, may be regarded as an aid to these controls, not as a substitute. As they depend upon the correct action in response to the warning, they are not effective unless combined with other measures. It may be necessary to include warnings in order to comply with national or local regulatory requirements.

TYPES OF CONTROL

Substitution of safer alternatives in procedures or materials is the first stage in the review of existing processes and procedures.

Isolation and enclosure of the process can be achieved by the use of physical barriers or by relocation of processes and/or facilities.

Local exhaust ventilation (LEV) is achieved by trapping the contaminant close to its source, and removing it directly by purpose-built ventilation prior to its entry into the breathing zone of the operator or the atmosphere.

LEV systems have four major parts, all of which must be efficiently maintained:

1. Hoods – the collection point
2. Ducting – to transport the contaminant away
3. Air purifying device – for example charcoal filters to prevent further pollution

4. Fan – the means of moving air through the system.

Efficiency of LEV systems is affected by draughts, capture hood design and dimensions, air velocity achieved and distance of capture point from the source. A major design consideration is that sucking air is very inefficient as an alternative to blowing it into a capture hood.

General or dilution ventilation uses natural air movement through open doors or assisted ventilation by roof fans or blowers to dilute the contaminant. It should only be considered if:

1. There is a small quantity of contaminant
2. The contaminant is produced uniformly in the area
3. The contaminant material is of low toxicity.

Good housekeeping lessens the likelihood of accidental contact with a contaminant. It includes measures to anticipate and handle spillages and leaks of materials, and minimising quantities in open use.

Reduced exposure time to a contaminant may be appropriate, provided that the possible harmful effect of the dose rate is taken into account. That is, high levels of exposure for short periods of time may be damaging.

Training should emphasise the importance of using the control measures provided, and give an explanation of the nature of the hazard which may be present together with the precautions which individuals need to take.

Personal protective clothing and equipment may be used where it is not possible to reduce the risk of injury sufficiently using the above control strategies. In that case, personal protective equipment must be used (See Section 8).

Welfare facilities allow workers to maintain good standards of personal hygiene, including regular washing and showering, and using appropriate clean protective clothing and equipment. The presence of adequate first-aid and emergency facilities minimises the effects of exposure to hazards.

Medical surveillance can detect early signs of ill-health. In some cases this can be carried out by supervisors trained to recognise the effects of exposure to workplace materials, or otherwise by the use of trained nursing and medical staff and facilities.

FAILURE OF CONTROLS

Once a control strategy has been devised and introduced, it is tempting to assume that there will be no further problems. A brief study of the possible ways in

Environmental Engineering Controls

which things can go wrong will show that there are two broad areas where problems can occur; in the introduction phase and as a result of change:

Inadequate initial design because of inappropriate choice of the type of control system, lack of consultations between designers, users and workers, failures to foresee future demands on the system, failure to consider consequences of introducing the system (e.g. increased noise).

Inadequate installation because of incompetence, or lack of adequate instructions or specifications.

Incorrect use may be due to lack of training, supervision or poor ergonomic design.

Inadequate maintenance resulting in blocked, damaged or removed parts, filters clogged or badly fitted, worn fans.

Failure to anticipate changes which may include the process itself, the materials used, the workers and supervisors concerned, work methods, the local environment including operator adjustments and additions, and regulatory changes such as exposure limits and LEV changes.

LEGAL REQUIREMENTS

The reader's attention is drawn to the comprehensive requirements for risk assessments to establish the need for environmental control measures. These include the Management of Health and Safety at Work Regulations 1992, Workplace (Health, Safety and Welfare) Regulations and the COSHH Regulations, as well as the Health and Safety at Work etc. Act 1974. Summaries of these can be found in Module D of this book.

SELF-ASSESSMENT QUESTIONS

1. Differentiate between local exhaust ventilation and dilution ventilation.

2. Identify the various environmental control strategies employed in your workplace.

REVISION

Eleven control strategies:

- Substitution
- Isolation
- Enclosure of the process
- Local exhaust ventilation
- General (dilution) ventilation
- Good housekeeping
- Reduced exposure time
- Training
- Personal protective clothing and equipment
- Welfare facilities
- Medical surveillance

Five causes of failure of controls:

- Inadequate initial design
- Inadequate installation
- Incorrect usage
- Inadequate maintenance
- Unanticipated change

Noise – Effects, Measurement and Control

INTRODUCTION

Noise enables us to communicate, and can create pleasure in the form of music and speech. However, exposure to excessive noise can damage hearing. Noise is usually defined as "unwanted sound", but in strict terms noise and sound are the same. Noise at work can be measured using a sound level meter. Sound is transmitted as waves in the air, travelling between the source and the hearer. The frequency of the waves is the **pitch** of the sound, and the amount of energy in the sound wave is the **amplitude**.

HOW THE EAR WORKS

Sound waves are collected by the **outer ear** and pass along the **auditory canal** about 2.5 cm to the **ear drum**. Changes in sound pressure cause the ear drum to move in proportion to the sound's intensity. On the inner side of the ear drum is the **middle ear** which is completely enclosed in bone. Sound is transmitted across the middle ear by three linked bones, the **ossicles** to the **oval window** of the **cochlea**, the organ of hearing which forms part of the **inner ear**. This is a spirally-wound tube, filled with fluid which vibrates in sympathy with the ossicles. Movement of the fluid causes stimulation of very small, sensitive cells with hairs protruding from them and rubbing upon a plate above them. The rubbing motion produces electrical impulses in the **hair cells** which are transmitted along the auditory nerve to the **brain** which then interprets the electrical impulses as perceived sound.

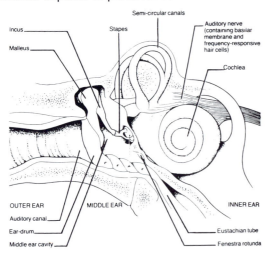

Hair cells sited nearest the middle ear are stimulated by high frequency sounds, and those sited at the tip of the cochlea are excited by low frequency. **Noise-induced hearing loss** occurs when the hair cells in a particular area become worn and no longer make contact with the plate above them. This process is not reversible, as the hair cells do not grow again once damaged.

HOW HEARING DAMAGE OCCURS

Excessive noise energy entering the system invokes a protection reflex, causing the flow of nerve impulses to be damped and as a result making the system less sensitive to low noise levels. This is known as **threshold shift**. From a single or short duration exposure, the resulting temporary threshold shift can affect hearing ability for some hours, but recovery then takes place. Repeated exposure can result in irreversible permanent threshold shift. The following damage can occur as a result of exposure to noise:

ACUTE EFFECTS:

1. **Acute acoustic trauma** from gunfire, explosions. Usually reversible, affects ear drum, ossicles.

2. **Temporary threshold shift** from short exposures, affecting the cochlea.

3. **Tinnitus** (ringing in the ears) results from intense stimulation of the auditory nerves, usually wears off within 24 hours.

CHRONIC EFFECTS:

1. **Permanent threshold shift** from long duration exposure, affects the cochlea and is irreversible.

2. **Noise-induced hearing loss** from (typically) long duration exposure, affects ability to hear human speech, irreversible, compensatable. It involves reduced hearing capability at the frequency of the noises that have caused the losses.

3. **Tinnitus**, as the acute form may become chronic without warning, often irreversibly.

Presbycusis is the term for hearing losses in older people. These have been thought to be due to changes due to aging in the middle ear ossicles, which causes a reduction in their ability to transmit higher-frequency vibrations.

MEASUREMENT OF NOISE

The range of human hearing from the quietest detectable sound to engine noise at the pain threshold is enormous, involving a linear scale of more than 100,000,000,000

Noise – Effects, Measurement and Control

units! Measuring sound intensity on such a scale would be clumsy, and so a method of compressing it is used internationally. The sound intensity or pressure is expressed on a logarithmic scale and measured in **bels**, although as a bel is too large for most purposes, the unit of measurement is the **decibel (dB)**. The decibel scale runs from 0 to 160 decibels (dB). A consequence of using the logarithm scale is that an increase of 3dB represents a doubling of the noise level. If two machines are measured when running separately at 90db each, the sound pressure level when they are both running together will not be 180dB, but 93dB. To establish noise levels on this scale, several different types of measurement are used.

Three weighting filter networks (A,B and C) are incorporated into sound level meters. They each adjust the reading given for different purposes, and the one most commonly used is the **'A' weighted dB**. This filter recognises the fact that the human ear is less sensitive to low frequencies, and the circuit attenuates or reduces very low frequencies to mimic the response of the human ear, and attach greater importance to the values obtained in the sensitive frequencies. Measurements taken using the A circuit are expressed in dB(A).

In most work places and most types of work, noise levels vary continuously. A measurement taken at a single moment in time is unlikely to be representative of exposure throughout the work period, yet this needs to be known as the damage done to hearing is related to the total amount of noise energy to which the ear is exposed.

A measure called L_{Eq} – the Continuous Equivalent Noise Level – is used to indicate an average value over a period which represents the same noise energy as the total output of the fluctuating real levels. L_{Eq} can be obtained directly from a sound level meter having an integrating circuit, which captures noise information at frequent timed intervals and recalculates the average value over a standard period, usually eight hours. It can also be calculated from a series of individual readings coupled with timings of the duration of each sound level, but this is laborious and relatively inaccurate.

Noise dose is a measure which expresses the amount of noise measured as a percentage, where 8 hours at a continuous noise level of 90 dB(A) is taken as 100%. If the work method and noise output is uniform, and the dose measured after 4 hours is 40%, then the likely 8-hour exposure will be less than 100%. However, if the dose reading after 2 hours is 60%, this will be an indication of an unacceptably high exposure.

$L_{EP,d}$ measures a worker's daily personal exposure to noise, expressed in dB(A). $L_{EP,w}$ is the measure of the worker's weekly average of the daily personal noise exposure, again expressed in dB(A).

Peak pressure is the highest pressure level reached by the sound wave, and assessments of this will be needed where there is exposure to impact or explosive noise. A meter capable of carrying out the measurement must be specially selected, because of the damping event of needle-based measuring which will consistently produce under-reading. A similar effect can be found in 'standard' electronic circuitry.

CONTROLLING NOISE

This can be achieved by:

Engineering controls – purchasing equipment which has low vibration and noise characteristics, and achieving designed solutions to noise problems including user quieter processes (e.g. presses instead of hammers), design dampers, mountings and couplings to be flexible, keeping sudden direction and velocity changes in pipework and ducts to a minimum. Operate rotating and reciprocating equipment as slowly as practicable.

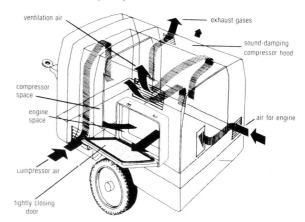

ventilation air — exhaust gases — sound-damping compressor hood — compressor space — engine space — air for engine — compressor air — tightly closing door

Orientation and location – moving the noise source away from the work area, or turning the machine around.

Enclosure – by surrounding the machine or other noise source with sound-absorbing material, but the effect is limited unless total enclosure is achieved.

Use of silencers – can suppress noise generated when air, gas or steam flow in pipes or are exhausted to atmosphere.

Lagging – can be used on pipes carrying steam or hot fluids as an alternative to enclosure.

Damping – can be achieved by fitting proprietary damping pads, stiffening ribs or by using double skin construction techniques.

Screens – are effective in reducing direct noise transmission.

Absorption treatment in the form of wall applications or ceiling panels; these must be designed for acoustic purposes to have significant effect.

Isolation of workers – in acoustically-quiet booths or control areas properly enclosed, coupled with scheduling of work periods to reduce dose will only be effective where there is little or no need for constant entry into areas with high noise levels. This is because even a short duration exposure to high sound pressure levels will exceed the permitted daily dose.

Personal protection – by the provision and wearing of ear muffs or plugs. This must be regarded as the last line of defence, and engineering controls should be considered in all cases. Areas where personal protective devices must be worn should be identified by signs, and adequate training should be given in the selection, fitting and use of the equipment, as well as the reasons for its use.

CHOICE OF HEARING PROTECTION

Hearing protection should be chosen to reduce the noise level at the wearer's ear to below the recommended limit for unprotected exposure. Selection cannot be based upon A-weighted measurement because effective protection will depend upon the ability of the protective device to attenuate (reduce) the sound energy actually arriving at the head position. Sound is a combination of many frequencies unless it is a pure tone, and it can happen that a particular noise against which protection is required has a frequency component which is not well handled by the 'usual' protection equipment. Therefore, a more detailed picture of the sound spectrum in question should be made before selection, checking the results obtained by **octave band analysis** against the sound-absorbing (attenuation) data supplied by the manufacturers of the products under consideration.

LEGAL REQUIREMENTS

Specific legislation on noise in all places of work has been introduced as the Noise at Work Regulations 1989, which replaces a previous Code and some specific Regulations. See Module D for a review of these measures, and also a summary of the Personal Protective Equipment at Work Regulations 1992, which require PPE to be selected according to criteria established in the risk assessment. New equipment provided must also conform to EC standards as a result of the activation of the PPE Product Directive.

SELF-ASSESSMENT QUESTIONS

1. Explain how noise-induced hearing loss is caused by noise at work.

2. What general controls are used in your workplace to reduce noise exposure?

REVISION

Two types of damage:

- ACUTE = acute acoustic trauma, temporary threshold shift, tinnitus
- CHRONIC = NIHL, PTS, tinnitus

Eleven noise control techniques:

- Engineering
- Orientation/Location

- Maintenance
- Enclosure
- Silencers
- Lagging
- Damping
- Screens
- Absorption treatment
- Isolation of workers
- Personal protective equipment

Personal Protective Equipment

INTRODUCTION

This Section discusses the methodology and practicalities of selecting and using personal protective equipment, or PPE. For a review of the legal requirements, please see Section 10 of Module D of this book.

PPE has a serious general limitation – it does not eliminate a hazard at source. If the PPE fails and the failure is not detected, the risk increases greatly. Where used, equipment must be appropriately selected and its use and condition monitored. Workers required to use it must be trained. For a PPE scheme to be effective, three elements must be considered:

Nature of the hazard – details are required before adequate selection can be made, such as the type of contaminant and its concentration.

Performance data for the PPE – manufacturer's information will be required concerning the ability of the PPE to protect against a particular hazard.

The acceptable level of exposure to the hazard – for some hazards the only acceptable exposure level is zero. Examples are work with carcinogens and protection of eyes against flying particles. Occupational Exposure Limits can be used, bearing in mind their limitations.

FACTORS AFFECTING USE

Once a decision to provide PPE has been made and a type selected, the following factors must be considered for their relevance. In a well-conceived programme for PPE, these constitute a basic checklist:

FIT – A good fit is required to ensure full protection. Some PPE is available only in a limited range of sizes and designs. Facial shape varies among different races, for example, and a facemask designed for the "average" Caucasian will be a poor fit on a Negroid face. Only limited fit adjustment is usually obtainable on the equipment itself.

PERIOD OF USE – It is necessary for the equipment to be worn whenever the hazard is present, so wearer acceptability is important. If the equipment is not acceptable for any reason, it is less likely to be worn, and will also affect task performance, concentration and stamina.

COMFORT – Although subjective, there is usually agreement among users about comfort, so permitting (limited) choice of equipment without compromising on protection standards will improve the likelihood of correct use.

MAINTENANCE – Equipment must be regularly cleaned, checked and maintained in serviceable condition.

TRAINING – Users and supervisors must know the limitations, correct use, how to achieve a good fit and the necessary maintenance for the equipment. This knowledge will be necessary to achieve the required standard of protection.

INTERFERENCE – Regard for the practicability of the item of equipment is needed in the work environment. Some eye protection interferes with peripheral vision, other types cannot easily be used with respirators. Correct selection can alleviate the problem, but full consideration must be given to the overall protection needs when selecting individual items, so that combined items of equipment may be employed, for example the "Airstream" helmet gives respiratory protection, and has fitted eye protection incorporated into the design.

MANAGEMENT COMMITMENT – the sine qua non of any safety programme, required especially in relation to PPE because it constitutes the last defence against hazards. Failure to comply with instructions concerning the wearing of it raises issues of industrial relations and corporate policy.

TYPES OF PPE

The types of PPE have different functions, including eye protection, hearing protection, respiratory protection, protection of the skin, and general protection in the form of protective clothing and safety belts, harnesses and lifelines.

HEARING PROTECTION: There are two main forms of hearing protection – objects placed in the ear canals to impede the passage of sound energy, and objects placed around the outer ear to restrict access of sound energy to the outer ear as well as the ear drum and middle and inner ear. It should be noted that neither of these forms of protection will prevent a certain amount of sound energy reaching the organ of hearing by means of bone conduction effects in the skull.

Ear plugs fit into the ear canal. They may be made from glass down, polyurethane foam or rubber, and are disposable. Some forms of re-usable plugs are available, but these are subject to hygiene problems unless great care is taken to clean them after use, and unless they are cast into the individual ear canal a good fit is unlikely to be achieved. Even though some plugs are available in different sizes, the correct size should only be determined by a qualified person. One difficulty is that in a reasonable proportion of people the ear canals are not the same size.

Long hair   Thick spectacle frames   Jewellery

Ear muffs consist of rigid cups which fit over the ears and are held in place by a head band. the cups generally have acoustic seals of polyurethane foam or a liquid-filled annular bag to obtain a tight fit. The cups are filled with sound-absorbing material. The fit is a function of the design of the cups, the type of seal and the tightness of the head band. the protective value of ear muffs may be lost almost entirely if objects such as hats or spectacles intrude under or past the annular seals.

RESPIRATORY PROTECTIVE EQUIPMENT (RPE): There are two broad categories; respirators, which purify the air by drawing it through a filter to remove contaminants, and breathing apparatus, which supplies clean air to the wearer from an uncontaminated external source. Most equipment will not provide total protection; a small amount of contaminant entry into the breathing zone is inevitable.

Five main types of respirator are available –

Filtering facepiece – consisting of a facepiece covering nose and mouth which is made of a filtering medium which removes the contaminant.

Half-mask respirator – which has a rubber or plastic facepiece covering the nose and mouth, and which carries one or more replaceable filter cartridges.

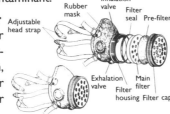

Full-face respirator – covering the eyes, nose and mouth and having replaceable filter canisters.

Powered air-purifying respirator – into which air is drawn through a filter via a pump and supplied under slight positive pressure which in turn prevents inward leakage of contaminant.

Powered visor respirator – which contains filters and/or fan mounted into a helmet so that purified air is blown down behind a protective visor.

Respirators do not provide ANY protection in oxygen-deficient atmospheres.

There are three main types of breathing apparatus available –

Fresh air hose apparatus – which supplies clean air from an uncontaminated source, pumped in by the breathing action of the wearer or by bellows.

Compressed airline apparatus – using flexible hosing delivering air to the wearer from a compressed air source. Filters in the airline are required to remove oxides of nitrogen and oil mist. Half-masks, full facepieces or hoods are used, the air pressure being reduced by valves.

Self-contained breathing apparatus – in which air is delivered to the wearer from a cylinder through a special mask. The complete unit is usually worn by the operative.

EYE PROTECTION: assessment of potential hazards to the eyes and the extent of the risks should be made in order to select equipment effectively. There are three types of eye protection commonly available:

Safety spectacles/glasses – which provide protection against low-energy projectiles such as metal swarf, but do not assist against dusts, are easily displaced and have no protective effect against high-energy impacts.

Safety goggles – to protect against high-energy projectiles and dusts, and are also available as protection against chemical and metal splashes with additional treatment. Disadvantages include a tendency to mist up inside, despite much design effort by manufacturers, lenses which scratch easily, limited vision for the wearer, lack of protection for the whole face, and high unit cost. Filters will be required for use against non-ionising radiation (See Section 9).

Face shields – offering high-energy projectile protection, also full-face protection and a range of special tints and filters to handle various types of radiation. Field of vision may be restricted. High initial cost of equipment is a disadvantage, although some visors allow easy and cheap replacement of shields. Weight can be a disadvantage, but this is compensated by relative freedom from misting up.

PROTECTIVE CLOTHING: provides body protection against a range of hazards, including heat and cold, radiation, impact damage and abrasions, water damage, chemical attack and impacts from vehicles (high-visibility wear).

Personal Protective Equipment

Head protection – is provided by two types of protectors; the safety helmet and the scalp protector (known also as the "bump cap") which is usually brimless. Their function is to provide protection against sun and rain, and against impact damage to the head. The ability of the scalp protector to protect against impacts is very limited, and its use is mainly to protect against bruising and bumps in confined spaces. It is not suitable for use as a substitute for a conventional safety helmet. Safety helmets have a useful life of about three years, which can be shortened by prolonged exposure to ultraviolet light and by repeated minor or major impact damage.

Protective outer clothing – is normally made of PVC material and often of high-visibility material to alert approaching traffic. PVC clothing can be uncomfortable to wear because of condensation, and vents are present in good designs. Alternatively, non-PVC fabric can be used which allows water vapour to escape, but this clothing is significantly more expensive.

Protective indoor clothing – such as overalls and coats are made of polycotton, and some makes are disposable. If overalls are supplied, arrangements for cleaning must be made to prevent unhygienic conditions developing if the clothing is worn in circumstances where oils or chemicals are handled. Failure to keep the clothing clean and changed regularly may result in dermatitis or skin cancer formation. Aprons and overtrousers should be fire-resistant, and trousers worn during cutting operations require protection in the form of ballistic nylon or similar material. Clothing may limit movement, and become entangled in machinery – careful selection of type and manufacturer is required, together with necessary training about its proper use. This may involve rules concerning the buttoning of coats in the vicinity of rotating machinery. Wearing anti-static clothing is of major importance in reducing static electricity effects – local rules should be strictly followed.

Gloves – must be carefully selected, taking account of use requirements such as comfort, degree of dexterity required, temperature protection offered and ability to grip in all conditions likely to be encountered, against considerations of cost and the hazards likely to be encountered by the wearer. For example, they can also become entangled in machinery. The main types of material and their features are:

MATERIAL GOOD FOR

Leather	Abrasion protection, heat resistance
PVC	Abrasion protection, water and limited chemical resistance
Rubber	Degreasing, paint spraying
Cloth/nylon latex coated	Hand grip
Latex	Electrical insulation work
Chain mail	Cut protection

Footwear – is designed to provide protection for the feet, especially for the toes, if material should drop or fall onto the feet. It should also protect against penetration from beneath the sole of the foot, be reasonably waterproof, provide a good grip, and be designed with reference to comfort. Steel toecaps are inflexible, and it is important to purchase the right size of footwear. Electrical insulation can be assisted by the correct footwear for the circumstances, and anti-static conducting shoes are essential where static effects need to be eliminated.

SKIN PROTECTION: Where protective clothing is not a practicable solution to a hazard, barrier creams may be used together with a hygiene routine before and after work periods. There are three types of barrier cream commonly found (water miscible, water repellent, special applications).

SAFETY BELTS AND HARNESSES: are not replacements for effective fall prevention practices. Only where the use of platforms or nets is impracticable is their use permissible. The functions of belts and harnesses are to limit the height of any fall, and to assist in rescues from confined spaces. In addition to comfort and freedom of movement, selection of this equipment must take into account the need to provide protection to the enclosed body against energy transfer in the event of a fall. Because of this, harnesses are generally preferable to belts except for applications where belts are required because of the movement needs of the work.

Belt and harness attachments to strong fixing points must be able to withstand the snatch load of the fall. A basic principle is to attach the securing lanyard to a fixing point as high as possible over the area of the work, so as to limit the fall distance. Similarly, a short lanyard should be provided. Equipment which has been involved in fall arresting should be thoroughly examined before further use, according to the manufacturer's instructions.

LEGAL REQUIREMENTS

The general duties Sections of the Health and Safety at Work etc. Act 1974 require a safe place of work to be provided together with safe systems of work, and these may involve use of PPE. General requirements for provision, use, maintenance and storage of PPE are contained in the Personal Protective Equipment at Work Regulations 1992. Specific Acts, Regulations and Orders also contain requirements for PPE, and should be consulted for particular applications.

SELF-ASSESSMENT QUESTIONS

1. An oxygen-deficient atmosphere has been established. Discuss the forms of RPE which should be used a) for continuous work in the area, b) for short term entry only.

2. List the protective equipment in your workplace. Can you identify its limitations and suggest alternative controls?

REVISION

Usage factors:

- Fit
- Period of use
- Comfort
- Maintenance
- Training
- Interference
- Management commitment

Types of PPE:

- Hearing protection
- RPE
- Eye/face protection
- Protective clothing
- Skin protection
- Safety belts and harnesses

Radiation

INTRODUCTION

Energy which is transmitted, emitted or absorbed as particles or in wave form is called **radiation**. Radiation is emitted by a variety of sources and appliances used in industry. It is also a natural feature of the environment. Transmission of radiation is the way in which radios, radar and microwaves work. The human body absorbs radiation readily from a wide variety of sources, mostly with adverse effects.

All types of electromagnetic radiation are similar in that they travel at the speed of light. Visible light is itself a form of radiation, having component wavelengths which fall between the infra-red and the ultraviolet portions of the spectrum. Essentially, there are two forms of radiation, ionising and non-ionising, which can be further subdivided.

RADIATION	EMITTED FROM	EXAMPLES OF HAZARDS
Radio frequency and microwaves	Communications, catering equipment, plastics welding	Heating of exposed body parts
Infra-red	Any hot material	Skin reddening, burns cataracts
Visible radiation	Visible light sources Laser beams	Heating, tissue destruction
Ultra-violet	Welding, some lasers carbon arcs	Sunburn, skin cancer, ozone
X-rays and other ionising radiations	Sources, radiography and X-ray machines	Burns, dermatitis, cancer, body cell damage

IONISATION AND RADIATION

All matter is made up of **elements**, which consist of similar **atoms**. These, the basic building blocks of nature, are made up of a **nucleus** containing **protons** and orbiting **electrons**.

Protons have a mass and a positive charge

Electrons have a negligible mass and a negative charge

Neutrons have a mass but no charge

If the number of electrons in an atom at a point in time is not equal to the number of protons, the atom has a nett positive charge, and becomes **ionised**. Ionising radiation is that which can produce ions by interacting with matter, including human cells, which leads to functional changes in body tissues. The energy of the radiation dislodges electrons from the cell's atoms, producing ion pairs, chemical free radicals and oxidation products. As body tissues are different in composition and form as well as in function, their response to ionisation is different. Some cells can repair radiation damage, others cannot. The cell's sensitivity to radiation is directly proportional to its reproductive capability.

IONISING RADIATION

Ionising radiations found in industry are alpha, beta and gamma, and X-rays. Alpha and beta particles are emitted from **radioactive** material at high speed and energy. Radioactive material is unstable, and changes its atomic arrangement so as to emit a steady but slowly diminishing stream of energy. **Alpha particles** are helium nuclei with two positive charges (protons), and thus are comparatively large and attractive to electrons. They have short ranges in dense materials, and can only just penetrate the skin. However, ingestion or inhalation of a source of alpha particles can place it close to vulnerable tissue, so essential organs can be destroyed. **Beta particles** are fast-moving electrons, smaller in mass than alpha particles but have longer range, so they can damage the body from outside it. They have greater penetrating power, but are less ionising.

Gamma rays have great penetrating power and are the result of excess energy leaving a disintegrating nucleus. Gamma radiation passing through a normal atom will sometimes force the loss of an electron, leaving the atom positively charged – an **ion**. This and the expelled electron are called an **ion pair**. Gamma rays are very similar in their effects to **X-rays**, which are produced by sudden acceleration or deceleration of a charged particle, usually when high-speed electrons strike a suitable target under controlled conditions. The electrical potential required to accelerate electrons to speeds where X-ray production will occur is a minimum of 15,000 volts. Equipment operating at voltages below this will not, therefore, be a source of X-rays. Conversely, there is a possibility of this form of radiation hazard being present at voltages higher than this. X-rays and gamma rays have high energy, and high penetration power through fairly dense material. In low density substances, including air, they may have long ranges.

Common sources in industry of ionising radiation are X-ray machines and isotopes used for nondestructive testing (NDT). They can also be found in laboratory work and in communications equipment.

NON-IONISING RADIATION

Generally, non-ionising radiations do not cause the ionisation of matter. Radiation of this type includes that in the electromagnetic spectrum between ultraviolet and radio waves, and also artificially-produced laser beams.

Ultraviolet radiation comes from the sun, and is also generated by equipment such as welding torches. Much of the natural ultraviolet in the atmosphere is filtered out by the ozone layer. Sufficient penetrates to cause sunburn and even blindness. Its effect is thermal and photochemical, producing burns and skin thickening, and eventually skin cancers. Electric arcs and ultraviolet lamps can produce a photochemical effect by absorption on the conjunctiva of the eyes, resulting in "arc-eye" and cataract formation.

Infra-red radiation is easily converted into heat, and exposure results in a thermal effect such as skin burning, and loss of body fluids. The eyes can be damaged, in the cornea and lens which may become opaque (cataract). Retinal damage may also occur if the radiation is focused, as in **laser** radiation. This is a concentrated beam of radiation, having principally thermal damage effects on the body.

Radiofrequency radiation is emitted by microwave transmitters including ovens and radar installations. The body tries to cool exposed parts by blood circulation. Organs where this is not effective are at risk, as for infra-red radiation. These include the eyes and reproductive organs. Where the heat of the absorbed microwave energy cannot be dispersed, the temperature will rise unless controlled by blood flow and

sweating to produce evaporation, convection and radiation. Induction heating of metals can cause burns when touched.

CONTROLS FOR IONISING RADIATION

The intensity of radiation depends upon the strength of the source, distance from it and the presence and type of shielding. Intensity will also depend upon the type of radiation emitted by the source. Radiation intensity is subject to the **inverse square law** – it is inversely proportional to the square of the distance from the source to the target. The dose received will also depend upon the duration of the exposure. These factors must be taken into account when devising the controls:

Elimination of exposure is the priority, to be achieved by restricting use and access, use of shielded enclosures and written procedures to cover:

- Use, operation, handling, transport, storage and disposal of known sources
- Identification of potential radiation sources
- Training of operators
- Identification of operating areas
- Monitoring of radiation levels around shielding
- Monitoring of personal exposure of individuals, by dosimeters
- Medical examinations for workers at prescribed intervals
- Hygiene practice in working areas
- Wearing of disposable protective clothing during work periods
- Clean-up practice
- Limiting of work periods when possible exposure could occur.

CONTROLS FOR NON-IONISING RADIATION

Protection against **ultraviolet radiation** is relatively simple; sunbathers have long known that anything opaque will absorb ultraviolet light. That emitted from industrial processes can be isolated by shielding and partitions, although plastic materials differ in their absorption abilities. Users of emitting equipment, such as welders, can protect themselves by the use of goggles and protective clothing – the latter to avoid "sunburn". Assistants often fail to appreciate the extent of their own exposure, and require similar protection.

Radiation

Visible light can, of course, be detected by the eye which has two protective control mechanisms of its own, the eyelids and the iris. These are normally sufficient, as the eyelid has a reaction time of 150 milliseconds. There are numerous sources of high-intensity light which could produce damage or damaging distraction, and sustained glare may also cause eye fatigue and headaches. Basic precautions include confinement of high-intensity sources, matt finishes to nearby paintwork, and provision of optically-correct protective glasses for outdoor workers in snow, sand or near large bodies of water.

Problems from **infra-red** radiation derive from thermal effects and include skin burning, sweating and loss of body salts leading to cramps, exhaustion and heat stroke. Clothing and gloves will protect the skin, but the hazard should be recognised so that effects can be minimised without recourse to personal protective equipment.

Controls for **laser** operations depend upon prevention of the beam from striking persons directly or by reflection. Effects will depend upon the power output of the laser, but even the smallest is a potential hazard if the beam is permitted to strike the body and especially the eye. Workers with lasers should know the potential for harm of the equipment they work with, and should be trained and authorised to do so. If the beam cannot be totally enclosed in a firing tube, eye protection should be worn suitable for the class of laser being operated (depends upon wavelength and intensity). Work areas should be marked so that inadvertent entry is not possible during operations. Laser targets require non-reflecting surfaces, and much care should be taken to ensure that this also applies to objects nearby which may reflect the laser beam. Toxic gases may be emitted by the target, so arrangements for ventilation should be considered. It is also necessary to ensure that the beam cannot be swung unintentionally during use, and that lasers are not left unattended whilst in use.

Equipment which produces **microwave** radiation can usually be shielded to protect the users. If size and function prohibits this, restrictions on entry and working near an energised microwave device will be needed. Metals, tools, flammable and explosive materials should not be left in the electromagnetic field generated by microwave equipment. Appropriate warning devices should be part of the controls for each such appliance. Commercially-available kitchen equipment is now subject to power restrictions and controls over the standard of seals to doors, but regular inspection and maintenance by manufacturers is required to ensure that it does not deteriorate with use and over time.

GENERAL STRATEGY FOR THE CONTROL OF EXPOSURE TO RADIATION

In addition to the previous specific controls, the following general principles must be observed:

1. Radiation should only be introduced to the workplace if there is a positive benefit.

2. Safety information must be obtained from suppliers about the type(s) of radiation emitted or likely to be emitted by their equipment.

3. There will be a requirement under the Management of Health and Safety at Work Regulations 1992 for written assessments of risks to be made, and noting the control measures in force. All those affected, including employees of other employers, the general public and the self-employed must be considered, and risks to them evaluated. They are then to be given necessary information about the risks and the controls. Relevant Modules and Sections of this book contain reviews of the requirements.

4. All sources of radiation must be clearly identified and marked.

5. Protective equipment (see 7 below) must be supplied and worn so as to protect routes of entry.

6. Safety procedures must be reviewed regularly.

7. Protective equipment provided must be suitable and appropriate, as required by relevant Regulations. It must be checked and maintained regularly.

8. A radiation protection adviser should be appointed with specific responsibilities to monitor and advise on use, precautions, controls and exposure

9. Emergency plans must cover the potential radiation emergency, as well as providing a control strategy for other emergencies which may threaten existing controls for radiation protection.

10. Written authorisation by permit should be used to account for all purchase/use, storage, transport and disposal of radioactive substances.

11. Workers should be classified by training and exposure period. Those potentially exposed to ionising radiation should be classed as 'persons especially at risk' under the Management of Health and Safety at Work Regulations 1992, Regulation 3(4)(b).

LEGAL REQUIREMENTS

General duties Sections of the Health and Safety at Work etc. Act 1974 apply to exposure to all forms of radiation, with its potential for harm. Specifically, the Ionising Radiations Regulations 1985 are the main control measure; together with duties concerning information provision contained in the Management of Health and Safety at Work Regulations 1992. See above for applicability of other measures, which are discussed further in Module D of this book.

SELF-ASSESSMENT QUESTIONS

1. Explain the difference between ionising and non-ionising radiations.

2. What control measures would be appropriate to ensure safety and health of employees using or working near a microwave oven?

REVISION

Two forms of radiation:

● Ionising – alpha, beta particles, gamma rays, X-rays

● Non-ionising – ultraviolet, infra-red, visible light, lasers, radiofrequency, microwaves

Ergonomics

INTRODUCTION

This Section offers a brief review only of a huge subject. The interested reader is invited to pursue further studies in the reference works at the end of this Module.

Accidents in industry and elsewhere often have a component of what we loosely describe as human error. It is a truism that 'to err is human', and we recognise this in one of the long-term objectives of health and safety management by attempting progressively to reduce the opportunities for people to make mistakes.

Consideration of human error is important because, even with automatic processes, people are still required to control, maintain and at certain points intervene. They also design and build the system, and can make mistakes at an early stage in the life of a process, system or machine. Understanding why people make errors is vital to the control of risks. The management of 'human factors' has been recognised as important, if difficult. Machines, procedures and processes must take human capabilities and fallibilities into account. Recent legislative initiatives, particularly from the EC, have incorporated the human factors approach. This can best be seen in the Manual Handling Operations Regulations 1992 and the Health and Safety (Display Screen Equipment) Regulations 1992.

The scope of 'human factors' includes:

- Perceptual, mental and physical capabilities of people
- Interaction of people with their jobs and working environment
- Influence of equipment and systems design on human performance
- Organisational characteristics which influence safety-related behaviour
- Social and inherited characteristics of the individual

For the individual at work, all of these factors are important, not only for physical health, safety and well-being, but also because they are less easy to control and influence than the 'hardware' aspects of working life – and thus the more easily neglected.

ERGONOMICS

Ergonomics is the applied study of the interaction between people, the objects and the environment around them. In the work environment, the objects include chairs, tables, machines and work stations. Ergonomics looks at more than just the design of chairs, though. A complete approach to the work environment is the aim, including making it easier to receive information from machines and interpret it correctly.

Careful ergonomic design improves the 'fit', and promotes occupational well-being. It also encourages employee satisfaction and efficiency. Ergonomics is concerned with applying scientific data on human mental and physical capabilities and performance to the design of workplaces, hardware and systems. Usually, the ergonomic design emphasis is on designing tools, equipment and work places so that the job fits the person rather than the other way around.

A combination of techniques is normally used. These include:

Work design – incorporating ergonomics into the design of tools, machines, work places and work methods. These are not mutually exclusive and a combination is often more appropriate to evolve the most appropriate solution.

Organisational arrangements – aimed at limiting the potentially harmful effects of physically demanding jobs on individuals. They may be concerned with selection and training, matching individual skills to job demands, job rotation methods and work breaks.

Studies carried out by the Swedish construction employment and insurance organisation Bygghälsan with employees and manufacturers have shown that considerable gains in productivity can be associated with optimising working conditions using ergonomic solutions. Recently, Bygghälsan has investigated musculo-skeletal disorders in the necks and shoulders of construction workers. A survey shows that almost half of all construction workers work more than 10 hours a week with their arms above shoulder level, leading to increased risk of problems in neck and shoulders. Sickness absences due to these problems increase for individuals over 30 in all trade categories in the industry, indicating that work-related problems in the neck and shoulders manifest themselves after 10-15 years of exposure.

It is impossible to eliminate totally the need to work with the hands above shoulder height, and Bygghälsan looked at alterations in the way the work was organised, work methods, improved equipment and techniques to make the work easier. One of the discoveries made was that the use of micro pauses – very short breaks – reduced muscular load and resulted in faster work. Screws were tightened into a beam at eye level, and those who took a 10-second pause after every other tightening did the work 12% faster than those who did not. Interestingly, the workers themselves thought it took more time to carry out the work with micro pauses than without.

Designing work and work equipment to suit the worker can reduce errors and ill-health – and accidents. Examples of problems which can benefit from ergonomic solutions are:

- Workstations which are uncomfortable for the operator
- Hand tools which impose strain on users
- Control switches and gauges which cannot be easily reached or read
- Jobs reported to be found excessively tiring

An ergonomic approach is now being used in the Manual Handling Operations Regulations 1992. It begins by forcing a review of the need to handle loads manually at all, and permits handling only following specific assessment of the risks associated with the task. The contrast with the former approach could not be more marked: we no longer train people to lift heavy weights as the sole solution to the problem.

ANTHROPOMETRY

Anthropometry is the measurement of the physical characteristics of the human body. People vary enormously in such basic characteristics as weight, height and physical strength. A car built for the 'average' passenger may require tall people to bend at uncomfortable angles, while smaller people may not be able to reach the controls. Designers use information on variations in size, reach and other physical dimensions to produce cars and other objects which most people can handle comfortably and conveniently. This has particular application in the design of workplaces and work equipment. Including these measurements in system design assists in the process of making man and machine more compatible, and produces considerable benefit to industry because of improvements in efficiency, quality and safety.

Anthropometry is concerned with measurements of body movement as well as static dimensions, and includes the discipline of **Biomechanics** which studies the forces involved in movements. Knowledge of all these factors is combined to make improvements in workplace design – seating, for example, and work equipment, where the knowledge assists in the design of appropriate machine guards.

THE 'AVERAGE' PERSON

Ergonomic designers usually try to meet the physical needs of the majority of a population. For any dimension, usually 10% of the population will fall outside the range used. That is, for example, the height of a workstation will only be wrong for the shortest 5% and the tallest 5% – the range is 5%-95%.

The designer will make deliberate compromises of this kind. It is not possible to produce a single size to fit everyone. Information about the design will enable purchasers to make informed decisions about equipment so that it meets the physical needs of the workforce.

SELF-ASSESSMENT QUESTIONS

1. Identify and list examples where ergonomic principles could be applied with benefit to your workplace. What improvements would you expect to result?

2. How can knowledge of ergonomics help in assessing manual handling risks?

Selected References

FOR ALL SECTIONS

Health and Safety at Work etc Act 1974

Management of Health and Safety at Work Regulations 1992

Workplace (Health, Safety and Welfare) Regulations 1992

Provision and Use of Work Equipment Regulations 1992

Factories Act 1961

Offices, Shops and Railway Premises Act 1963

and associated Approved Codes of Practice and Guidance on Regulations

OCCUPATIONAL EXPOSURE LIMITS

The Control of Substances Hazardous to Health Regulations 1988 (COSHH)

The Ionising Radiations Regulations 1985

The Control of Asbestos at Work Regulations 1987

The Control of Lead at Work Regulations 1980

HSE guidance

EH40/year Occupational exposure limits

EH64 Occupational exposure limits – criteria document summaries

American Conference of Governmental & Industrial Hygienists Incorporated

Threshold limit values (for chemical substances and physical agents) and biological exposure indices

ENVIRONMENTAL MONITORING

The Control of Substances Hazardous to Health Regulations 1988 (COSHH)

The Ionising Radiations Regulations 1985

The Control of Asbestos at Work Regulations 1987

The Control of Lead at Work Regulations 1980

and associated Approved Codes of Practice

HSE guidance

EH42 Monitoring strategies for toxic substances

EH10 Asbestos – control limits and measurement of airborne dust concentrations

ENVIRONMENTAL ENGINEERING CONTROLS

The Control of Substances Hazardous to Health Regulations 1988 (COSHH)

The Ionising Radiations Regulations 1985

The Control of Asbestos at Work Regulations 1987

The Control of Lead at Work Regulations 1980

HSE guidance

EH22 Ventilation of the Workplace

EH44 Dust in the workplace – general principles of protection

HS(G)37 An introduction to local exhaust ventilation

HS(G)54 The maintenance, examination and testing of local exhaust ventilation

NOISE

Noise at Work Regulations 1989

Construction Plant and Equipment (Harmonisation of Noise Emission Standards) Regulations 1985

HSE guidance

Noise at Work Guides 1 and 2 – legal duties of employers, designers, manufacturers, importers and suppliers to prevent damage to hearing

Noise at Work Guides 3-8 – Noise assessment, information and control

PERSONAL PROTECTIVE EQUIPMENT

Personal Protective Equipment at Work Regulations 1992

Control of Substances Hazardous to Health Regulations 1988

Noise at Work Regulations 1989

Construction (Head Protection) Regulations 1989

Control of Asbestos at Work Regulations 1987

Control of Lead at Work Regulations 1980

Ionising Radiations Regulations 1985

HSE guidance

L25 Personal Protective Equipment – guidance on Regulations

EH41 RPE for use against asbestos

EH53 RPE for use against airborne radioactivity

HS(G)53 RPE – a practical guide for users

Head Protection Regulations – advice and guidance (leaflet)

Module B Section 11

British Standards

BS1397   Specifications for industrial safety belts, harnesses and safety lanyards
BS1870   Specifications for safety footwear
BS2092   Specification for eye protectors for industrial/non-industrial use
BS4275   Recommendations for selection, use and maintenance of RPE
BS5240   Specification for general purpose industrial safety helmets
BS6344   Industrial hearing protectors
BS7028   Selection and maintenance of eye protection for industrial and other uses

RADIATION

Ionising Radiations Regulations 1985
Radiological Protection Act 1970

Guidance (HSC/E)

COP 16   The protection of persons against ionising radiations arising from any work activity: The Ionising Radiations Regulations 1985
COP 23   Exposure to radon
EH53   RPE for use against airborne radioactivity
GS41   Radiation safety in underwater radiography
HS(G)49   The examination and testing of portable radiation instruments for external radiations

British Standard

BS1542   Specification for equipment for eye, face and neck protection against non-ionising radiation during welding and similar operations

ERGONOMICS

The Manual Handling Operations Regulations 1992
The Health and Safety (Display Screen Equipment) Regulations 1992
Workplace (Health, Safety and Welfare) Regulations 1992
Provision and Use of Work Equipment Regulations 1992
Personal Protective Equipment at Work Regulations 1992

Guidance (HSE)

L22   Work Equipment – guidance on Regulations
L23   Manual Handling – guidance on Regulations
L25   Personal Protective Equipment – guidance on Regulations
L26   Display Screen Equipment – guidance on Regulations

HS(G)48   Human factors in industrial safety
HS(G)38   Lighting at work

Working with VDU's – leaflet

Other sources

Nicholson AJ and Ridd J (eds) 'Health, safety and ergonomics' 1988 Butterworths ISBN 0 40 802386 4

Module C:
Safety Management Techniques

INTRODUCTION

Accidents are the direct results of unsafe activities and conditions, both of which are able to be controlled by management. Management is responsible for the creation and maintenance of the working environment and tasks, into which workers must fit and inter-react. Control of this environment has been discussed in previous Modules. Control of workers and their behaviour is more difficult. They have to be given information, and the knowledge that accidents are not inevitable but are caused. They need training to develop skills and recognise the need to comply with and develop safe systems of work, and to report and correct unsafe conditions and practices. Their safety awareness and attitudes require constant improvement, and the social environment of the workplace must be one which fosters good safety and health practices and conditions, not one which discourages them.

A primary requirement of management is to appreciate the need to concentrate on the nature of the accident phenomenon rather than its outcome, the injury or damage/loss. Also, there is a need for awareness that the primary cause of an accident is not necessarily the most important feature; secondary causes, usually system failures, will persist unless action is taken. A 'simple' fall from a ladder may be dismissed as 'carelessness', but this label may hide other significant factors, such as lack of training, maintenance, adequate job planning and instruction, and no safe system of work. These topics will be discussed in this Module.

Three statements must be adopted by management in order to achieve success by planning rather than by chance:

1. Accidents are caused
2. Steps must be taken to prevent them and achieve the three objectives below
3. Accidents will continue to happen if these steps are not taken.

ACCIDENT PREVENTION OBJECTIVES

1. **MORAL** objectives derive from the concept that a duty of reasonable care is owed to others. Greater awareness of the quality of life at work has focussed popular attention on the ability of employers to handle a wide variety of issues, previously seen only as marginally relevant to the business enterprise. Environmental affairs, pollution, product safety and other matters are now commonly discussed, and there is a growing belief that it is simply morally unacceptable to put the safety and health of others inside or outside the workplace at risk, for profit or otherwise. Physical pain and hardship resulting from death and disability is impossible to quantify. Moral obligations are now more in the minds of employers than ever before.

A dimension of the moral objective is **morale**, which also interlinks with the following two objectives. Worker morale is strengthened by active participation in accident prevention programmes, and weakened following accidents. Adverse publicity affects the fortunes of the organisation both internally in this way and externally, as public confidence may weaken local community ties, market position, market share and reputation generally.

2. **LEGAL** objectives are given in statute law, which details steps to be taken and carries the threat of prosecution or other enforcement action as a consequence of failure to comply. Civil law enables injured workers and others to gain compensation as a result of breach of statutory duties or because a reasonable standard of care was not provided under the circumstances.

3. **ECONOMIC** objectives are to ensure the continuing financial health of a business and avoid the costs associated with accidents. These include monetary loss to employers, community and society from worker injuries, damage to property and work interruptions. Some, but not all of these costs are insurable and are known as **direct** costs. Increased premiums will be a consequence of claims, so an increase in overheads is predictable following accidents. **Indirect** costs include uninsured property damage, delays, overtime costs, management time spent on investigations, and decreased output from the replacement(s).

BASIC TERMS

An **accident** is an incident plus its consequences; the end product of a sequence of events or actions resulting in an undesired consequence (injury, property damage, interruption, delay). The **incident** is that sequence of events or actions. An incident does not necessarily have a definable start or finish. (Think about a road bulk tanker overturning and spilling its contents onto the road and down drains. Can you say when the incident started or finished?) An **injury** is thus a consequence of failure – but not the only possible one. It has been estimated that hundreds of incidents occur in industry for every one which causes injury or loss, but all have the

Accident Prevention

potential to do so. That is why it is important to look at all incidents as sources of information on what is going wrong. Relying on injury records only allows a look at a minority of incidents which happened to result in a serious injury consequence.

An **accident** can be defined more fully as 'an undesired event, which results in physical harm and/or property damage, usually resulting from contact with a source of energy above the ability of the body or structure to withstand it.' The idea of energy transfer as part of the definition of an accident is a relatively recent one, which helps our understanding of the accident process.

Hazard means the inherent property or ability of something to cause harm – the potential to interrupt or interfere with a process or person, which is or may be causally related to an accident, by itself or with other variables. Hazards may arise from interacting or influencing components, for example two chemicals interacting to produce a third.

Risk is the chance or probability of loss, an evaluation of the potential for failure. It is easy to confuse the terms 'hazard' and 'risk', and many writers have done so. The terms are often incorrectly used, sometimes interchanged. A simple way to remember the difference is that 'hazard' describes potential for harm, risk is the likelihood that harm will result in the particular situation or circumstances. Another way of defining risk is that it is the probability that a hazard will result in an accident with definable consequences. In a wider sense, we can look at 'risk' as the product of the severity of the consequences of any failure and the likelihood of that failure occurring. Thus, an event with a low probability of occurrence but a high severity can be compared against an event likely to happen relatively often but with a comparatively trivial consequence. Comparisons between risks can be made using simple numerical formulae (See Section 3).

ACCIDENT CAUSES

Immediate or **primary** causes of accidents are often grouped into unsafe acts and unsafe conditions. This is convenient, but can be misleading as accidents typically include both groups of causes at some stage in the chain of causation.

UNSAFE ACTS can include:

- Working without authority
- Failure to warn others of danger
- Leaving equipment in a dangerous condition
- Using equipment at the wrong speed

- Disconnecting safety devices such as guards
- Using defective equipment
- Using equipment the wrong way or for wrong tasks
- Failure to use or wear personal protective equipment
- Bad loading of vehicles
- Failure to lift loads correctly
- Being in an unauthorised place
- Unauthorised servicing and maintaining moving equipment
- Horseplay
- Smoking in areas where this is not allowed
- Drinking alcohol or taking drugs.

UNSAFE CONDITIONS can include:

- Inadequate or missing guards to moving machine parts
- Defective tools, equipment
- Inadequate fire warning systems
- Fire hazards
- Ineffective housekeeping
- Hazardous atmospheric conditions
- Excessive noise
- Exposure to radiation
- Inadequate illumination or ventilation.

These are all deviations from required safe practice, but they must be seen as the symptoms of more basic underlying **indirect** or **secondary** causes which allow them to exist and persist.

INDIRECT CAUSES include:

- Management system pressures
  - financial restrictions
  - lack of commitment
  - lack of policy
  - lack of standards
  - lack of knowledge and information
  - restricted training and selection for tasks.

- Social pressures
  - group attitudes
  - trade customs
  - tradition
  - society attitudes to risk-taking
  - 'acceptable' behaviour in the workplace.

PRINCIPLES OF ACCIDENT PREVENTION

In summary, these are: the systematic use of techniques to identify and remove hazards, the control of risks which remain, and the use of techniques to influence behaviour and encourage safe attitudes. They are the primary responsibility of management, and are discussed in the next Sections.

Section 2 covers techniques of health and safety management, including the central role of the management organisation and policy needed to put them into practice, Section 3 discusses the assessment of risk, and Section 4 the details of successful written health and safety policies. Detailed treatments of practical aspects are contained in Modules A and B.

SELF-ASSESSMENT QUESTIONS

1. Explain the difference between hazard and risk, using examples from your workplace.

2. Write a short summary of an accident with which you are familiar, and list the immediate cause(s) and the indirect cause(s).

REVISION

Three accident prevention objectives:

- Moral/morale
- Legal
- Economic

Definitions: accident, incident, hazard, risk

Accident causes are immediate or indirect, involve unsafe acts and/or conditions

Three principles of accident prevention:

- Use techniques to identify and remove hazards
- Use techniques to assess and control remaining risks
- Use techniques to influence behaviour and attitudes

Techniques of Health and Safety Management

INTRODUCTION

Safety management is concerned with and achieved by all the techniques which promote health and safety. Some have been described in other Modules, some will be considered in Sections later in this Module. Safety management is also concerned with influencing human behaviour, and with limiting the opportunities for mistakes to be made which would result in harm or loss. To do this, safety management must take into account the ways in which people fail (fail to do what is expected of them and/or what is safe). Limiting risks requires elimination or control of hazards, and assessment of risks which remain.

OBJECTIVES

The practical objectives of safety management are:

Gaining support from all concerned for the health and safety effort
Motivation, education and training so that all may recognise and correct hazards
Achieving hazard control by design and purchasing
Operation of a suitable inspection programme to provide feedback (See Section 10)
To ensure hazard control principles form part of supervisory training
Devising and introducing controls based on risk assessments
Compliance with Regulations and standards.

To achieve these objectives, a safety policy statement is required. The design and other considerations for this policy are discussed in Section 4.

BENEFITS

Successful safety management can lead to substantial cost savings, as well as a good accident record. Some companies have become well-known for the success of their safety management system. Du Pont claims several of its plants with more than 1,000 employees have run for more than ten years without recording a lost-time injury accident. Du Pont has ten principles of safety management which are worthy of study:

1. All injuries and occupational illnesses are preventable.

2. Management is directly responsible for doing this, with each level accountable to theone above and responsible for the level below.

3. Safety is a condition of employment, and is as important to the company as production, quality or cost control.

4. Training is required in order to sustain safety knowledge, and includes establishing procedures and safety performance standards for each job.

5. Safety audits and inspections must be carried out.

6. Deficiencies must be corrected promptly, by modifications, changing procedures, improved training and/or consistent and constructive disciplining.

7. All unsafe practices, incidents and injury accidents will be investigated.

8. Safety away from work is as important as safety at work.

9. Accident prevention is cost-effective; the highest cost is human suffering.

10. People are the most critical element in the health and safety programme. Employees must be actively involved, and complement management responsibility by making suggestions for improvements.

KEY ELEMENTS

The key elements of successful health and safety management are:

- Policy
- Organising
- Planning and implementing
- Measuring performance
- Reviewing performance and auditing

The reader interested in a complete treatment of this topic is referred to the HSE publication HS(G)65 'Successful health and safety management'.

POLICY

Successful health and safety management demands comprehensive health and safety policies which are effectively implemented and which are considered in all business practice and decision making. Also, some jurisdictions require written safety policy statements to be created by all employers, except for the smallest organisations. This has been the case in the United Kingdom since April 1975, and is simply a reflection in the law of what has been known for many years – written policies are the centrepieces of good health and safety management. They insist, persuade, explain, and assign responsibilities. An essential requirement for management involvement at all levels is to define health and safety responsibility in detail within the written document, and then to check at intervals that the responsibility has been adequately discharged. This process leads to **ownership** of the health and safety programme, and it is based on the principle of accountability discussed in Section 4.

ORGANISING

To make the health and safety policy effective, the staff must be actively involved and committed. Organisations which achieve high standards in health and safety create and sustain a 'culture' which motivates and involves all members of the organisation in the control of risks. They establish, operate and maintain structures and systems which are intended to:

- Secure **control** – by ensuring managers lead by example

- Encourage **co-operation** – both of employees and their trade union safety representatives

- Secure effective **communication** – by providing information about hazards, risks and preventive measures

- Ensure **competence** – by assessing the skills needed to carry out all tasks safely, and then by providing the means to ensure that all employees (including temporary ones) are adequately instructed and trained. The Management of Health and Safety at Work Regulations 1992 encourage this by requiring employers to recruit, select, place, transfer and train on the basis of assessments and capabilities, and to ensure that appropriate channels are open for access to specialist advice when required.

PLANNING AND IMPLEMENTING

Planning ensures that health and safety efforts really work. Success in health and safety relies on the establishment, operation and maintenance of planning systems which:

- **Identify** objectives and targets which are attainable and relevant

- **Set performance standards** for management, and for the control of risks which are based on hazard identification and risk assessment, and which take legal requirements as the accepted minimum standard

- **Consider and control** risks both to employees and to others who may be affected by the organisation's activities, products and services

- **Ensure documentation** of all performance standards.

Organisations which plan and control in this way can expect fewer claims, reduced insurance costs, less absenteeism, higher production, improved quality and lower operating costs.

MONITORING

Just like finance, production or sales, health and safety has to be monitored to establish the degree of success. For this to happen, two types of monitoring system need to be operated. These are:

- **Active** monitoring systems – are intended to measure the achievement of objectives and specified standards before things go wrong. This involves regular inspection and checking to ensure that standards are being implemented and that management controls are working properly.

- **Reactive** monitoring systems are intended to collect and analyse information about failures in health and safety performance, when things do go wrong. This will involve learning from mistakes, whether they result in accidents, ill-health, property damage incidents or 'near misses'.

Information from both active and reactive monitoring systems should be used to identify situations that create risks and enable something to be done about them. Priority should be given to where the risks are greatest. The information should then be referred to people within the organisation who have the authority to take any necessary remedial action, and also to effect any organisational and policy changes which may be necessary.

REVIEWING AND AUDITING PERFORMANCE

Auditing enables management to ensure that their policy is being carried out and that it is having the desired effect. Auditing complements the monitoring programme. Economic auditing of a company is well established as a tool to ensure economic stability and it has been shown that similar systematic evaluations of performance in health and safety has equal benefits. An audit is not the same as an inspection. Essentially, the audit assesses the organisation's ability to meet its own standards on a wide front, rather than providing a 'snapshot' of a particular site or premises.

The two main objectives of an audit are:

- To ensure that standards achieved conform as closely as possible to the objectives set out in the organisation's safety policy

Techniques of Health and Safety Management

- To provide information to justify continuation of the same strategy, or a change of course.

The best health and safety audit systems are capable of identifying deviations from agreed standards, analyse events leading to these deviations and highlight good practice. They look especially at the 'software' elements of health and safety such as systems of work, management practices, instruction, training and supervision as well as the more traditional 'hardware' elements such as machinery guarding and the use of personal protective equipment.

PEOPLE PROBLEMS

In all control measures, reliance is placed upon **human behaviour** to carry out the solutions, and a major task of health and safety management is to assure safe behaviour by motivation, education, training, and the creation of work patterns and structures which enable safe behaviour to be practised. In a major study in 1977, it was found that supervisors in the construction industry gave a variety of reasons for their inactivity on health and safety matters. In order of frequency, the most common responses were:

Resource limitations
Seen as outside the boundaries of their duties
Acceptance of hazards as inevitable
Influences of the social climate
Industry tradition
Lack of technical competence
Incompatible demands upon their time
Reliance upon the worker to take care
Lack of authority
Lack of information

It can be seen that much of the inactivity of **supervision** can be corrected by establishing a favourable environment, with clear responsibilities given and accountability practised, together with necessary training in the complex nature of the accident phenomenon and in solutions to health and safety problems. Supervisors' and workers' attitudes to safety generally reflect their perception of the attitudes of the employer.

Attempts to motivate the individual meet with success when persuasion rather than compulsion is used to achieve safety objectives. Consultation with workers, through representatives such as Trade Unions, and locally through the formation of **safety committees**, is generally a successful strategy provided that an adequate role is given to those being consulted.

Benefits of safety committees include the involvement of the workforce, encouragement to accept safety standards and rules, help in arriving at practicable solutions to problems, and exposure of hazards which may not be apparent to management.

At an **individual** level, an appreciation of some of the more common reasons why people fail to carry out tasks safely is useful for management. People may actually decide to act wrongly – sometimes in the belief that non-observance of safety rules is expected of them by managers, supervisors or work colleagues. Few deliberately decide to injure themselves, but a deliberate decision to err may be made following a poor estimate of the risk of injury in an activity. Others err because of traps – design of equipment so that the correct action cannot be taken because of physical inability to do so. Valves placed out of reach, dials too far away to be properly read, poor ergonomic design of work stations all contribute to the likelihood that mistakes will be made. Under these circumstances, it is not surprising that motivational aids such as posters are found to be relatively ineffective in producing safe behaviour.

PRIORITIES IN PREVENTION

Basic principles should be observed in setting up strategies for control and management of health and safety at work. These are:

- If possible, avoid a risk altogether by eliminating the hazard

- Tackle risks at source – avoid the quick temporary fix or putting up a sign where a better physical control could be used (quieten machines rather than provide personal protective equipment or putting up warning signs)

- Adapt work to the individual when designing work areas and selecting methods of work

- Use technology to improve conditions

- Give priority to protection for the whole workplace rather than to individuals (protect a roof edge rather than supply safety harnesses)

- Ensure everyone understands what they have to do to be safe and healthy at work

- Make sure health and safety management is accepted by everyone, and that it applies to all aspects of an organisation's activities.

Techniques of Health and Safety Management

SELF-ASSESSMENT QUESTIONS

1. What are the advantages of a works safety committee? Are there any disadvantages?

2. How is the success of your safety management programme evaluated? Are positive measures used as well as negative ones?

REVISION

Six objectives of health and safety management:

- To gain support from all concerned
- Motivation, education and training
- Achieving hazard control
- Operating inspection programme
- Devising and introducing risk controls
- Compliance with Regulations and standards

Five key elements of health and safety management:

- Policy
- Organising
- Planning and implementing
- Measuring performance
- Review and auditing

Two monitoring systems:

- Active
- Reactive

Risk Assessment

INTRODUCTION

There are at least two senses in which risk assessment has been carried out unconsciously over a long period. Firstly, we all make assessments many times each day of the relative likelihood of undesirable consequences arising from our actions in particular circumstances. Whether to cross a road by the lights or take a chance in the traffic is one such. In making a judgment we evaluate the chance of injury and also the likely severity.

The second sense of risk assessment is based on the requirement for the employer under the Health and Safety at Work etc. Act 1974 in many of its Sections to take 'reasonably practicable' precautions in various areas to safeguard employees. In doing this, we make a balanced judgment about the extent of the risk and its consequences against the time, trouble and cost of the steps needed to remove or reduce it. If the cost is 'grossly disproportionate', we are able to say that the steps are not reasonably practicable. Thus, in a very real sense, risk assessments have been carried out at least since 1974 in the UK.

The difference between these assessments and those now required by the Management of Health and Safety at Work Regulations 1992 (MHSWR) is that the significant results are to be recorded by most employers, and information based on them is to be given to employees in a much more specific way than before.

BENEFITS

Risk assessment is done to enable control measures to be devised. We need to have an idea of the relative importance of risks and to know as much about them as we can in order to take decisions on controls which are both appropriate and cost-effective.

TYPES OF RISK ASSESSMENT

There are two major types of risk assessment, which are not mutually exclusive. The type which produces a objective probability estimate based upon known risk information applied to the circumstances being considered – this is a **quantitative** risk assessment. The second type is subjective, based upon personal judgment backed by generalised data on risk – the **qualitative** assessment. Except in cases of specially high risk, public concern, or where the severity consequences of an accident are felt to be large and widespread, qualitative risk assessments are much simpler to make and

adequate for their purpose. The legal requirements refer to this type of assessment unless the occasion demands more rigorous methods to be used.

Where there is similarity of activities, hazards and risks associated with them although present in different physical areas, or workplaces, a general risk assessment can be made which covers their basic features. This is known as a 'generic' or 'model' assessment, and should be included in the safety policy document. There are likely to be situations in specific areas or on specific occasions when such an assessment will not be sufficiently detailed, and those circumstances should be indicated in the policy to alert to the need to take further action. There may also be work situations where hazards associated with particular situations will be so unique that a special assessment must be made every time the work is done. Examples include demolition, the erection of steel structures and asbestos removal.

CONTENTS OF RISK ASSESSMENTS

Significant findings are to be recorded where five or more are employed by an employer. According to the Approved Code of Practice, the record should contain a statement of significant hazards identified, the control measures in place and the extent to which they control the risk (cross-reference can be made to manuals and other documents), and the population exposed to the risk.

As the Regulations require review of assessments in specified circumstances, and as it will be prudent to review for changed circumstances over time, it would be prudent to include in the documentation a note of the date the assessment was made and the date for the next regular review. Similarly, it would be wise to include a note to employees reminding them of their duties under Regulation 12 of MHSWR to inform the employer of any circumstances which might indicate a short-coming in the assessment.

HAZARD EVALUATION

The hazards to be identified are those associated with machinery, equipment, tools, procedures, tasks, processes and the physical aspects of the plant and premises – everything. Evaluation of the hazards is achieved by assembling information from those familiar with the hazards, such as insurance companies, professional societies, government departments and agencies, manufacturers, consultants, trade unions, old inspection reports both internal and externally-produced, accident reports and standards.

Some hazards may not be readily identifiable, and there are techniques which can be applied to assist in this respect. These include inductive analysis, which predicts failures – Failure Modes and Effects Analysis (FMEA) is one of these, job safety analysis (JSA) is another. Inductive analysis assumes failure has occurred and then examines ways in which this could have happened by using logic diagrams. This is time-consuming, and therefore expensive, but it is extremely thorough. MORT (Management Oversight and Risk Tree) Analysis is an example which is not difficult to use.

RANKING HAZARDS BY RISK

Ranking produces a priority list of hazards to be controlled, on a 'worst first' basis. It takes account of the consequence (likely severity) and the probability of the event occurring. Estimation of the first is easier than the second, as data may not be available for all hazards. Estimates derived from experience can be used. It is possible to carry out ranking using a simple formula, where risk = severity estimate x probability estimate. These estimates can be given any values, as long as they are consistently used. The simplest set of values offers a 16-point scale:

SEVERITY RATING OF HAZARD　　　**VALUE**

Catastrophic – imminent danger exists, hazard capable of causing death and illness on a wide scale.　　　1

Critical – Hazard can result in serious illness, severe injury, property and equipment damage.　　　2

Marginal – Hazard can cause illness, injury or equipment damage, but the results would not be expected to be serious.　　　3

Negligible – Hazard will not result in serious injury or illness, remote possibility of damage beyond minor first-aid case.　　　4

PROBABILITY RATING OF HAZARD　VALUE

Probable – Likely to occur immediately or shortly　　　1

Reasonably probable – Probably will occur in time　　　2

Remote – May occur in time　　　3

Extremely remote – Unlikely to occur　　　4

NB. These categories are capable of much further refinement and words like 'time' can be defined,

increasing if necessary the number of categories. Many organisations increase the categories to take account of numbers exposed to the hazard, and duration of exposure. However, the more precise the definitions, the more it will be necessary to possess accurate predictive data.

DECISION-MAKING

This process requires information to be available on alternatives to the hazard, also other methods of controlling the risk. Factors which will influence decision-making are training, possibility of replacement of equipment or plant, modification possibilities and the cost of solutions proposed. **Cost benefit analysis** will be used formally or informally at this point.

Cost benefit analysis requires a value to be placed upon the costs of improvements suggested or decided upon. These will include the cost of reducing the risk, eliminating the hazard, any capital expenditure needed, and any on-going costs applicable. An estimate of the pay-back period will be needed. Decisions on action to be taken are often based on this – a 3-5 year period is often associated with health and safety improvements. Using these techniques will direct organisation resources to where they are most needed.

INTRODUCTION OF CORRECTIVE AND PREVENTATIVE MEASURES

The reader's attention is drawn to the principles contained in the Approved Code of Practice to the MHSW Regulations, which are summarised at the end of Section 2. These principles are complemented by the further concept of **progressive risk reduction**, which encourages the setting of improving objectives of risk reduction year by year.

It is necessary to remember that some corrective measures are better than others, and some are very ineffective indeed as controls. The **safety precedence sequence** shows the order of effectiveness of measures:

Hazard elimination (for example, use of alternatives, design improvements, change of process)

Substitution (for example, replacement of a chemical with one with less risk)

Use of barriers:
　　Isolation (removes hazard from the worker, puts hazard in a box)

Risk Assessment

Segregation (removes worker from the hazard, puts worker in a box)

Use of procedures:
Limiting exposure time, dilution of exposure
Safe systems of work, depending upon human response

Use of warning systems:
Signs, instructions, labels, which depend upon human response

Use of personal protective equipment:
Depends upon human response, used as a sole measure only when all other options have been exhausted – PPE is the last resort

NB. See Module B, Section 8 for discussion of personal protective equipment. Note that Regulations now require it should only be used if there is no immediately feasible way to control the risk by more effective means, and as a temporary measure pending installation of more effective solutions. Disadvantages of PPE include – interference with ability to carry out the task; PPE may fail and expose the worker to the full effect of the hazard; continued use may mask presence of the hazard and result in no further preventive action being taken.

MONITORING

Risk assessments must be checked to ensure their validity, or when reports indicate that they may no longer be valid. It is important to remember that fresh assessments will be required when the risks themselves change as conditions change, and also when new situations and conditions are encountered for the first time. Other information relevant to risk assessments will come from monitoring by way of inspection (see Section 10), air quality monitoring and other measurements (See Module B Section 5) and medical surveillance.

HEALTH SURVEILLANCE

Risk assessments will identify circumstances where health surveillance will be appropriate. Requirements for such surveillance now extend beyond exposure to substances hazardous to health. Generally, there will be a need if there is an identifiable disease or health condition related to the work, there is a valid technique for its identification, there is a likelihood that the disease or condition may occur as a result of the work, and that the surveillance will protect further the health of employees. Examples where these conditions may apply are vibration white finger and forms of work-related upper limb disorders (WRULDS).

INFORMATION TO OTHERS

Regulations 9 and 10 of the MHSWR require co-operation and sharing of information between employers sharing or acting as host at a workplace. Risk assessments will form the basis of that information. In formal circumstances of contract, there may be a contractual or legal requirement to exchange risk assessment data. The information given to others about risks must include the health and safety measures in place to address them, and be sufficient to allow the other employers to identify anyone nominated to help with emergency evacuation.

SELF-ASSESSMENT QUESTIONS

1. What information will be needed before an assessment is carried out?

2. Thinking about qualitative and quantitative risk assessments, explain the difference with examples showing where it would be appropriate for each to be used.

REVISION

Two types of risk assessment:

- Qualitative
- Quantitative

Risk = Severity x probability

Five contents of risk assessments:

- Hazard details
- Applicable standards
- Evaluation of risks
- Preventive measures
- Review dates/feedback details

INTRODUCTION

Without active support, any attempt at organised accident prevention will be useless – or even worse than useless, since there may be an illusion that health and safety matters are under control, resulting in complacency. Avoidance of accidents requires a sustained, integrated effort from all departments, managers, supervisors and workers in an organisation. Only management can provide the authority to ensure this activity is co-ordinated, directed and funded. Its influence will be seen in the policy made, the amount of scrutiny given to it, and the ways in which violations are handled.

THE SAFETY POLICY

The most effective means of demonstrating management commitment and support is by issuing a safety policy statement, signed and dated by the most senior member of the management team, and then ensuring that the requirements of the policy are actively carried out by managers, supervisors and workers. Lack of firm management direction of this kind encourages the belief that "safety is some-one else's business".

Section 2(3) of the Health and Safety at Work etc. Act 1974 requires written safety policy statements to be created by all employers, except for the smallest organisations (with fewer than five employees). This is simply a reflection in the law of what has been known for many years – written policies are the centrepieces of safety management. They insist, persuade, explain, and assign responsibilities. An essential requirement for management involvement at all levels is to define health and safety responsibility in detail within the written document, and then to check at intervals that the responsibility has been adequately discharged. This is **accountability** – the primary key to management action. It is not the same thing as responsibility; accountability is responsibility that is evaluated and measured, possibly during appraisal sessions.

SAFETY POLICY CONTENTS

The statement itself is an expression of management intention, and as such does not need to contain detail. What is usually referred to as 'the safety policy' will contain this statement, together with the details of the organisation (responsibilities at each level within the operation) and the arrangements (how health and safety will be managed, in detail). It is important to distinguish between a safety policy and a safety manual – these are not the same, but are often found combined. The safety policy will refer to the manual for details on technical points. The main problem is that the likelihood of a document being read is proportional to its length and complexity. Current opinion is that safety policies should be shorter rather than longer, and accompanied by explanatory manuals.

The safety policy should provide all concerned with concise details of the organisation's health and safety goals, objectives and the means of achieving them, including the assignment of responsibility and detailed arrangements for each workplace. Organisations with several workplaces in different locations find it convenient to express management philosophy in an overall statement, leaving a detailed statement and policy to be written and issued at local level.

Risk assessments as required by the Management of Health and Safety at Work Regulations 1992 will form a part of the safety policy document, as they form the basis for deciding on the control measures – the arrangements – and detailing the responsibilities within the organisation. The Regulations require that the significant findings of the risk assessments are to be recorded where there are five or more employees. The practicality of attaching risk assessments in full to the safety policy document is such that it will best be done by including the findings of generic assessments (See Section 3) in detail, and referring to circumstances when these will need modification and where specific assessments will be required on each occasion the risk is faced.

ORGANISATION AND ARRANGEMENTS

The main headings of matters which should be detailed in the safety policy are:

Responsibilities of management and supervisors at all levels

Duties of workers, including statutory and organisation rules

Role and functions of health and safety professional staff

Allocation of finance for health and safety

Systems used to monitor safety performance (not just injury recording)

Identification of main hazards likely to be encountered by the workforce

Generic risk assessments – significant findings

The Safety Policy

Any circumstances when specific risk assessments will be required

Arrangements (or cross-references) for dealing with them

Safety training policy, details of arrangements

Design safety

Fire arrangements

Occupational health facilities, including first-aid

Environmental monitoring policy and arrangements

Purchasing policy (e.g. on safety, noise, chemicals)

Methods of reporting accidents and incidents

Methods used to investigate accidents and incidents

Arrangements for the use of contractors

Personal protective equipment policy, requirements, availability

Worker consultation arrangements (e.g. safety committees)

OTHER CONSIDERATIONS

Safety policies, as written statements of the intentions of management, acquire a quasi-legal status. Among other things, they serve as a record of the standard of care provided by the management. This offers a useful method of evaluating an organisation in terms of health and safety, especially because the standards, beliefs and commitments are on view. Therefore, the document will be useful when evaluating contractors (see Section 8). Others may wish to use it for different purposes – to gauge the record of a possible business partner, and, importantly, for the purpose of establishing an employer's self-assessed standard of care as a prelude to making a civil claim for damages. Claimants may be able to use any extravagant wording or undertakings to further claims, as they can demonstrate what should have been available or shown in their case.

Revision of safety policies should be done at regular intervals, to ensure that organisation and arrangements are still applicable to the organisation's needs. After changes in structure, senior personnel, work arrangements, processes or premises the hazards and risks may change. After incidents and accidents, one of the objectives of the investigation (see Section 9) will be to check that the arrangements in force had anticipated the circumstances and foreseen the causes of the accident. If they had not, then a change to the policy will be required.

Circulation/distribution of the policy is important – it must be read and understood by all those affected by it. How this is best achieved will be a matter for discussion. In some organisations, a complete copy is given to each employee. In others, a shortened version is given out, with a full copy available for inspection at each workplace. Members of the management team should be familiar with the complete document. If it is likely to be revised frequently, a loose-leaf format will be advisable. This is especially true if the names and contact phone numbers of staff are printed in it; these change frequently.

THE FUNCTION OF THE SAFETY POLICY

Safety policy statements serve as the bridge between management health and safety activities and the legal system, because a requirement exists in the **statute law** for a policy to be written, and because the policy can be used in **civil** claims. Generally, legal compliance for its own sake is the poorest of reasons for accident prevention – apart from the moral considerations, the financial losses associated with poor safety performance outweigh likely penalties for breaches of statutory duty.

SELF-ASSESSMENT QUESTIONS

1. Why are safety policies useful in the management of health and safety? Why would one consisting of only one page be of no value?

2. Obtain a copy of your organisation's safety policy. Can you think of any improvements which might usefully be made?

REVISION

Safety policies contain:

- A general statement of management philosophy
- Details of the organisation (responsibilities)
- Risk assessment significant findings

- The arrangements in force to control the risks
- Signature of the most senior member of management
- The date of last revision

INTRODUCTION

It has been estimated that at least a quarter of all fatal accidents at work involve failures in systems of work – the way things are done. A safe system of work is a formal procedure which results from a systematic examination of a task in order to identify all the hazards and assess the risks, and which identifies safe methods of work to ensure that the hazards are eliminated or the remaining risks are minimised.

Many hazards are clearly recognisable and can be overcome by separating people from them physically e.g. using guarding on machinery (see Module A Section 1). There may be circumstances where hazards cannot be eliminated in this way, and elements of risk remain associated with the task. **Where the risk assessment indicates this is the case, a safe system of work will be required.**

Some examples where safe systems will be required as part of the controls are:

- Cleaning and maintenance operations
- Changes to normal procedures, including layout, materials and methods
- Working alone or away from the workplace and its facilities
- Breakdowns and emergencies
- Control of the activities of contractors in the workplace
- Vehicle loading, unloading and movements

DEVELOPING SAFE SYSTEMS

Some safe systems can be verbal only – where instructions are given on the hazards and the means of overcoming them, for short duration tasks. These instructions must be given by supervisors or managers – leaving workers to devise their own method of work is not a safe system of work. The law requires a suitable and sufficient risk assessment be made of all the risks to which employees and others who may be affected by them are exposed. Although some of these assessments can be carried out using a relatively unstructured approach, a more formal analysis can be used to develop a safe system of work. Sometimes these may be carried out as a matter of policy, with the task broken down into stages and the precautions associated with each written into the final document. This can be used for training new workers in the required method of work. The technique is known as **job safety analysis**.

For all safe systems, there are five basic steps necessary in producing them:

- Assessment of the task
- Hazard identification and risk assessment
- Identification of safe methods
- Implementing the system
- Monitoring the system

TASK ASSESSMENT

All aspects of the task must be looked at, and should be put in writing to ensure nothing is overlooked. This should be done by supervision in conjunction with workers involved, to ensure that assumptions of supervisors about methods of work are not confounded by reality. Account must be taken of **what is used** – the plant and substances, potential failures of machinery, substances used, electrical needs of the task – **sources of errors** – possible human failures, short cuts, emergency work – **where the task is carried out** – the working environment and its demands for protection, and **how the task is carried out** – procedures, potential failures in work methods, frequency of the task, training needs.

HAZARD IDENTIFICATION AND RISK ASSESSMENT

Against a list of the elements of the task, associated hazards can be clearly identified, and a risk assessment can be made. A more complete review of risk assessment can be found in the preceding Section. Where hazards cannot be eliminated and risks reduced, procedures to ensure a safe method of work should be devised.

DEFINITION OF SAFE METHODS

The chosen method can be explained orally as already mentioned. Simple written methods can be established, or a more formal method known as a permit-to-work system. All of these involve setting up the task and any authorisation necessary; planning of job sequences; specification of the approved safe working methods including the means of getting to and from the task area if appropriate; conditions which must be verified before work starts – atmospheric tests, machinery lockout; and dismantling/disposal of equipment or waste at the end of the task.

Safe Systems of Work

IMPLEMENTING THE SYSTEM

There must be adequate communication if the safe system of work is to be successful. The details should be understood by everyone who has to work with it, and it must be carried out on each occasion. It is important that everyone appreciates the need for the system its place in the accident prevention programme.

Supervisors must know that their duties include devising and maintaining safe systems of work, and making sure they are put into operation, and revised where necessary to take account of changed conditions or accident experience. Training is required for all concerned, to include the necessary skills, awareness of the system and the hazards which it is aimed to eliminate by the use of safe procedures. Part of every safe system should be the requirement to stop work when a problem appears which is not covered by the system, and not to resume until a safe solution has been found.

MONITORING THE SYSTEM

Effective monitoring requires that regular checks are made to make sure that the system is still appropriate for the needs of the task, and that it is being fully complied with. Checking only after accidents is not an acceptable form of monitoring. Simple questions are required – do workers continue to find the system workable? Are procedures laid down being carried out? Are the procedures still effective? Have there been any changes which require a revision of the system? A system devised as above which is not followed is **not** a safe system of work – the reasons must be found and rectified. Safe systems of work are associates of, not substitutes for, the stronger prevention techniques of design, guarding and other methods which aim to eliminate the possibility of human failure.

LEGAL REQUIREMENTS

Section 2(2)(a) of the Health and Safety at Work etc. Act 1974 requires the provision and maintenance of safe systems of work that are, so far as is reasonably practicable, safe and without risks to health. Under the employer's general duty of care at common law, a failure to so gives rise to a claim based on the allegation of the employer's negligence. Specific legislation requires the use of formal safe systems, called permits-to-work, either directly or by implication as a means of compliance (for example, confined space entry).

Further requirements for safe systems of work following upon risk assessments are contained in the Management of Health and Safety at Work Regulations 1992, which also place duties on employees to follow the systems and procedures set up for their protection following risk assessments.

SELF-ASSESSMENT QUESTIONS

1. Can you think of tasks in your workplace which require correct steps to be taken whilst carrying them out to ensure worker safety? Are they covered by written documentation? Can you justify those which are only given orally?

2. Write out a safe system of work for moving filing cabinets between offices.

REVISION

Five steps in devising a safe system:

- Assessment of the task
- Identification of the hazards and assessment of the risks
- Definition of safe method
- Implementation of the safe system
- Monitoring the system

Formal safe systems may be required for:

- Cleaning tasks
- Maintenance work
- Changes to routine
- Working alone
- Working away from normal environment
- Breakdowns
- Emergencies
- Contractors
- Vehicle operations

INTRODUCTION

Training for health and safety is not an end in itself, it is a means to an end. Talking in general terms to employees about the need to be safe is not training; workers and management alike need to be told what to do for their own health and safety and that of others, as well as what is required by statute. A knowledge of what constitutes safe behaviour in a variety of different occupational situations is not inherited but must be acquired, either by trial and error or from a reputable source of expertise. Trial and error methods are likely to extract too high a price in modern industry, where the consequences of forced and unforced errors may be very serious, even catastrophic.

Experience and research also shows that knowledge of safe behaviour patterns, gained by instruction, films, videos, posters and booklets, does not guarantee that safe behaviour will be obtained from individuals. Training is therefore never a substitute for safe and healthy working conditions, and good design of plant and equipment. Because humans are fallible, the need is to lessen the opportunities for mistakes and unsafe behaviour to occur, and to minimise the consequences when it does.

Safety training may be a part of other training in work or organisational procedures for reasons of time or cost, and there is merit in combined training as it serves to emphasise the need to regard health and safety generally as an integral part of good business management.

Training in any subject requires the presence of three necessary conditions before it commences: the active commitment, support and interest of management, necessary finance and organisation to provide the opportunity for learning to take place, and the availability of suitable expertise in the subject. The support of management demonstrates the presence of an environment into which the trained person can return and exercise new skills and knowledge. The management team also demonstrates support by setting good examples; it is pointless to train workers to obey safety rules if supervisors are known to ignore them.

Trainers must not only be knowledgeable in their subject, but also be qualified to answer questions on the practical application of the knowledge in the working environment, which will include a familiarity with organisational work practices, procedures and rules.

TRAINING NEEDS

New employees are known to be more likely to have accidents than those who have had time to recognise the hazards of the workplace. Formal health and safety training is now required by law to form part of the induction programme. Training must also take place when job conditions change and result in exposure to new or increased risks. It must be repeated periodically where appropriate, and be adapted to any new circumstances. No health and safety training can take place outside working hours.

There may also be opportunities for self-instruction, perhaps using modern technology to assist, for example computer-based interactive learning programmes. The key points which should be covered in induction training are:

● Review and discussion of the organisation's overall safety programme or policy, and the policy relating to the work activities of the newcomers

● Safety philosophy; safety is as important as production or any other organisational activity, accidents have causes and can be prevented, prevention is the primary responsibility of management, each employee has a personal responsibility for his or her own safety and that of others

● Local, national and organisational health and safety rules or regulations will be enforced, and those violating them may be subject to some form of discipline

● The health and safety role of supervisors and other members of the management team includes taking action on and giving advice about potential problems, and they are to be consulted if there are any questions about the health and safety aspects of work

● Where required, the wearing or use of personal protective equipment is not a matter for individual choice or decision – its use is a condition of employment

● In the event of any injury, no matter how trivial it may appear, workers must seek first-aid or medical treatment and notify their supervisor immediately

● Fire and emergency procedure(s)

● Welfare and amenity provision

● Arrangements for joint consultation with workers and their representatives should be made known to all newcomers. A joint approach to health and safety problems, and the regular reviewing of work practices, procedures, systems and written documentation is an essential part of a good health and safety programme (but joint or balanced participation should not be used as a method of removing or passing off the prime responsibility of management at all levels to manage health and safety at least as efficiently as other aspects of the organisation).

Job-specific training should include skills training, explanations of applicable safety regulations and organisation rules and procedures, a demonstration of any personal protective equipment which may be required and provided for the work (including demonstration of correct fit, method and circumstances of use and cleaning procedures), the handing-over of any documentation required, such as permit-to-work documents, safety booklets and chemical information sheets. There should also be a review of applicable aspects of emergency and evacuation procedures. This training may be carried out by a supervisor, but it should be properly planned and organised by the use of checklists.

Supervisory and general management training at all levels is necessary to ensure that responsibilities are known and the organisation's policy is carried out. Management failures which have come to light following investigations into disasters, plant accidents and other health and safety incidents have been concentrated in the following areas:

- Lack of awareness of the safety systems, including their own job requirements for health and safety

- Failure to enforce health and safety rules adequately or at all

- Failure to inspect and correct unhealthy or unsafe conditions

- Failure to inform or train workers adequately

- Failure to promote health and safety awareness by participation in discussions, motivating workers and setting an example.

It is not sufficient simply to tell supervisors they are responsible and accountable for health and safety; they must be told the extent of the responsibilities and how they can discharge them. Key points to cover in the training of supervisors and managers are:

- The organisation's safety programme and policy
- Legal framework and duties of the organisation, its management and the workforce
- Specific laws and rules applicable to the work area
- Safety inspection techniques and requirements
- Causation and consequences of accidents
- Basic accident prevention techniques
- Disciplinary procedures and their application
- Control of hazards likely to be present in the work area, including machinery safety, fire, materials

handling, hazards of special equipment related to the industry, use of personal protective equipment
- Techniques for motivating employees to recognise and respond to organisational goals in health and safety.

Senior managers should be given essentially the same information, as this gives them a full appreciation of the tasks of subordinates, makes them more aware of standards of success and failure, and equips them to make cost-beneficial decisions on health and safety budgeting.

External assessment of the training given to management at all levels is desirable. This can be done by training to the appropriate syllabi of national or international professional organisations, and encouraging those trained to take the relevant examination. In the UK this is the National Certificate in Occupational Safety and Health, an examination including a premises inspection which is administered by the National Examinations Board in Occupational Safety and Health (NEBOSH).

Examples of specialised training needs

FIRST-AID TRAINING has been proved to be a significant factor in accident prevention, as well as obviously beneficial in aiding accident victims. People trained in first-aid are more safety conscious and are less likely to have accidents. Project FACTS – First Aid Community Training for Safety – was launched in 1970 in Ontario, Canada, and arranged for the training of 5,500 people in industry, schools and among the public in the community of Orillia, Ontario. Results showed that first-aid training throughout the workforce can cut accident rates by up to 30%. A second project based on industry alone found that employees not trained in first-aid had accident rates double that of employees of similar age, sex, job and employer.

Apart from the above, and legal requirement which may exist, the need for first-aid training in the workplace depends upon a number of factors. These include the nature of the work and the hazards, what medical services are available in the workplace, the number of employees and the location of the workplace relative to external medical assistance. Shift working may also be taken into account, also the ratio of trained persons present to the total number of workers.

DRIVER TRAINING and certification may be a requirement for particular classes of vehicle, according to national or local regulations in force, and in these cases the detail of the training programme may be defined in law. A common cause of death at work and away from work is the traffic accident. As the loss of a key worker can have a severe impact upon the viability of an organisation, training for all workers who drive should be considered. Defensive driver training has been found effective and cost-beneficial in reducing numbers of traffic accidents, and has been extended to members of workers' families, particularly those entitled to drive employer's vehicles.

FIRE TRAINING to the extent that all workers should know the action to take when fire alarms sound, should be given to all employees, and included in induction training. Knowledge of particular emergency plans and how to tackle fires with equipment available may be given in specific training at the workplace. At whatever point the training is given, the following key points should be covered:

- Evacuation plan for the building in case of fire, including assembly point(s)
- How to use fire fighting appliances provided
- How to use other protective equipment, including sprinkler and other protection systems, and the need for fire doors to be unobstructed
- How to raise the alarm and operate the alarm system from call points
- Workplace smoking rules
- Housekeeping practices which could permit fire start and spread if not carried out, e.g. waste disposal, flammable liquid handling rules
- Any special fire hazards peculiar to the workplace.

Fire training should be accompanied by practices, including regular fire drills and evacuation procedures. No exceptions should be permitted during these.

Reinforcement training will be required at appropriate intervals, which will depend upon observation of the workforce (training needs assessment), on the complexity of the information needed to be held by the worker, the amount of practice required and the opportunity for practice in the normal working environment. Assessment will also be needed of the likely severity of the consequences of behaviour which does not correspond with training objectives when required to do so. If it is absolutely vital that only certain actions be taken in response to plant emergencies, then more frequent refresher training will be needed to ensure that routines are always familiar to those required to operate them.

LEGAL REQUIREMENTS

Many specific pieces of health and safety legislation contain requirements to provide training for employees engaged in certain tasks, and the most commonly applicable of these have been selected for discussion in Module D. A general duty for all employees to be trained (and provided with information, instruction and supervision in addition) as necessary to ensure their health and safety so far as is reasonably practicable is contained in Section 2(2)(c) of the Health and Safety at Work etc. Act 1974, and more specific requirements are contained in the Management of Health and Safety at work Regulations 1992 (See Module D Section 5).

SELF-ASSESSMENT QUESTIONS

1. What topics should be covered in induction safety training to provide necessary knowledge and skills for fire prevention?

2. "Managers don't need to know about safety rules – they are not at risk". Discuss.

REVISION

Five types of health and safety training needs:

- New employee induction
- Job-specific for new starters

- Supervision and management
- Specialised
- Reinforcement

Maintenance

INTRODUCTION

Maintenance can be defined as work carried out in order to keep or restore every facility (part of a workplace, building and contents) to an acceptable standard. It is not simply a matter of repair; some mechanical problems can be avoided if preventative action is taken in good time. Health and safety issues are important in maintenance, because statistics show the maintenance worker to be at greater risk of accidents and injury. This is partly because these workers are exposed to more hazards than others, and partly because there are pressures of time and money upon the completion of the maintenance task as quickly as possible. Less thought is given to the special needs for health and safety of the maintenance task than is given to routine tasks which are easier to identify, plan and control.

MAINTENANCE POLICY

Maintenance standards are a matter for the organisation to determine; there must be a cost balance between intervention with normal operations by planned maintenance and the acceptance of losses because of breakdowns or other failures. From the health and safety aspect, however, defects requiring maintenance attention which have led or could lead to increased risk for the workforce should receive a high priority. Linking inspections with maintenance can be useful, so that work areas and equipment are checked regularly for present and possible future defects. Some plant items may be subject to statutory maintenance requirements, and the manufacturer's instructions in this respect should be complied with as well.

PREVENTATIVE MAINTENANCE

The need for maintenance of any piece of equipment should have been anticipated in its design. Lubrication and cleaning will still be required for machinery, but the tasks can be made safer by consideration of maintenance requirements at an early stage of design, and later at installation. A written plan for preventative maintenance is required, which documents the actions to be taken, how often this needs to be done, all health and safety matters associated, the training required (if any) before maintenance work can be done, and any special operational procedures such as permits-to-work and locking-off required.

BREAKDOWN MAINTENANCE

The number of failures which require this will be reduced by planned maintenance, but many circumstances (such as severe weather conditions) can arise which require workers to carry out tasks beyond their normal work experience, and/or which are more than usually hazardous by their nature. Records of all breakdowns should be kept, to influence future planned maintenance policy revisions, safety training and design. (See also Module A Sections on safe design and machine guarding for discussion of the need for maintenance considerations to be recognised at the design stage).

Health and safety aspects. Maintenance accidents are caused by one or more of the following factors:

- Lack of perception of risk by managers/supervisors, often because of lack of necessary training.

- Unsafe or no system of work devised, for example no permit-to-work system in operation, no facility to lock off machinery and electrics before work starts and until work has finished.

- No co-ordination between workers, or communication with other supervisors or managers.

- Lack of perception of risk by workers, including failure to wear protective clothing or equipment, no co-ordination.

- Inadequacy of design, installation, siting of plant and equipment.

- Use of contractors who are inadequately briefed on health and safety aspects, not selected on health and safety as well as cost grounds (see Section 8).

MAINTENANCE CONTROL TO MINIMISE HAZARDS

The safe operation of maintenance systems requires steps to be taken to control the above factors. These steps can be divided into the following phases:

Planning – identification of the need for planned maintenance and arranging a schedule for this to meet any statutory requirements. A partial list of items for consideration includes air receivers and all pressure vessels, boilers, lifting equipment, electrical tools and machinery, fire and other emergency equipment, and structural items under wear, such as floor coverings.

Evaluation – the hazards associated with each maintenance task must be listed and the risks of each considered (frequency of the task and possible consequences of failure to carry it out correctly). The tasks can then be graded and the appropriate degree of management control applied to each.

Control – the control(s) for each task will take the above factors into account, and will include any necessary review of design and installation, training, introduction of written safe working procedures to minimise risk (which is known as a safe system of work, as discussed in Section 5), allocation of supervisory responsibilities, and necessary allocation of finances. Review of the activities of contractors engaged in maintenance work will be required in addition (see Section 8).

Monitoring – random checks, safety audits and inspections, and the analysis of any reported accidents for cause which might trigger a review of procedures constitute necessary monitoring to ensure the control system is fully up to date. The introduction of any change

in the workplace may have maintenance implications and should therefore be included in the monitoring process.

LEGAL REQUIREMENTS

Many pieces of health and safety legislation contain both general and specific requirements to maintain premises, plant and equipment. The general duty is contained in Section 2(2)(d) of the Health and Safety at Work etc. Act 1974, and more specific requirements are contained in the Management of Health and Safety at work Regulations 1992 (See Module D Section 5). Other requirements are contained in the Workplace (Health, Safety and Welfare) Regulations 1992, the Personal Protective Equipment at Work Regulations 1992 and the Provision and Use of Work Equipment Regulations 1992, all of which are summarised in Module D.

SELF-ASSESSMENT QUESTIONS

1. Outline the maintenance system which would lead to the rapid repair of a non-working machine guard.

2. How are repairs made to the roof of your workplace? Can the maintenance system be made safer?

REVISION

Maintenance accidents are due to:

- Inadequate design
- Lack of perception of risk
- Lack of a safe system of work
- Communication failures
- Failure to brief and supervise contractors

Maintenance accidents can be prevented by:

- Planning
- Evaluation
- Controls
- Monitoring

Management and Control of Contractors

INTRODUCTION

Anyone entering premises for the purposes of carrying out specialised work for the client, owner or occupier must be regarded as a 'contractor' – to whom duties are owed, and indeed who owes duties with regard to health and safety matters. Because of this, the same kinds of control measures must be applied to all who work on premises – caterers, window cleaners, agency staff, equipment repairers and servicers – the list is long and must be written down as a first control task.

Analysis of investigations into accidents shows that financial pressures, whether real or perceived, are nearly always present. The making and acceptance of the low bid in competitive tendering is often at the expense of health and safety standards. Other major factors include a transient labour force which never gets properly or fully trained, the small size of most contracting companies which claim not to be aware of legislation or safe practices, the inherent danger of the work and work conditions, pressure of work, and poor management awareness of the need for safety management.

A CONTROL STRATEGY

There are six parts to a successful control strategy. The extent to which each part is relevant will depend upon the degree of risk and the nature of the work to be contracted. The parts are:

- Identification of suitable bidders
- Identification of hazards within the specification
- Checking of (health and safety aspects of) bids and selection of contractor
- Contractor agrees to be subject to client's rules
- Control of the contractor on site
- Checking after completion of contract

IDENTIFICATION OF SUITABLE CONTRACTORS

It is clearly necessary to work out a system aimed at ensuring that a contractor with knowledge of safety standards and a record of putting them into practice is selected for the work.

1. Each contractor wishing to enter an 'approved list' should be asked to provide his safety policy. Arrangements will be required for vetting these for adequacy.

2. A prequalification questionnaire should be completed by each contractor, providing necessary information about his policy on health and safety, including details of responsibility, experience, safe systems of work and training standards.

3. At this stage, it should be possible to identify contractors for approval, but feedback will be required to identify any who do not in practice conform to their own stated standards. This means that the list will require regular scrutiny and updating.

SPECIFICATION

A checklist should be followed which will give a pointer to most if not all of the common health and safety problems which may arise during the work. These should be communicated to the contractor in the specification before the bid is made, and the received bid checked against them to ensure that proper provision is being made for the control of risk and that the contractor has identified the hazards. Suitable headings for the checklist include:

1. Special hazards and applicable national or local Regulations and Codes of practice (asbestos, noise, permits to work)

2. Training required for the contractor's employees

3. Safe access/egress to, from, on the site, and to places of work **within** the site

4. Electrical and artificial lighting requirements

5. Manual/mechanical lifting

6. Buried and overhead services

7. Fire protection

8. Occupational health risks, including noise

9. Confined space entry

10. First-aid/emergency rescue

11. Welfare amenities

12. Safe storage of chemicals

13. Personal protective equipment

14. Documentation and notifications

15. Insurance and special terms and conditions of the contract

The process of drawing-up the specification should include appropriate consultations with the workforce to ensure that they are fully aware of the proposed work and have a chance to comment on it as it affects their own health and safety.

CHECKING THE BIDS

When the bids are returned, it should be possible to distinguish the potentially competent at this stage. An 'approved list' of contractors, scrutinised at intervals, can save the need for carrying out a complete selection process as described on every occasion.

SAFETY RULES

A basic principle of control is that as much as possible should be set down in detail in the contract. An important condition should be that the contractor agrees to abide by all the provisions of the client's safety policy which may affect his employees or the work, including compliance with site health and safety rules. Exchange of information and summaries of risk assessments is a requirement of Regulations 9 and 10 of the Management of Health and Safety at Work Regulations 1992, where construction or any contract workers interact with the employer's undertaking.

Often, the contractor may delegate the performance of all or part of the contract to other sub-contractors. In these cases it is essential to ensure that the sub-contractors are as aware as the original contractor of the site rules and safety policy. A condition which can be attached to the contract is that the contractor undertakes to inform any sub-contractors of all safety requirements, to incorporate observance of them as a requirement of any future sub-contract, and to require the sub-contractor to do likewise if he in turn sub-contracts any work.

Written orders containing detailed terms and conditions such as the above should be the basis of the contract and should be acknowledged by the contractor before work starts. The loan of tools and equipment by the client should be avoided unless part of the original contractual arrangement.

Areas of concern which should be covered by general site rules and within the client's safety policy should be communicated to the contractor in the form of site rules. They include:

- Materials storage, handling, disposal
- Use of equipment which could cause fires
- Noise and vibration
- Scaffolding and ladders, access

- Cartridge-powered fixing tools
- Welding equipment – and use of client's electricity supply
- Lifting equipment – certificated, adequate
- Competency of all plant operators
- Vehicles on site – speed, condition, parking restrictions
- Use of lasers, ionising radiation
- Power tools – voltage requirements
- Machinery brought on site
- Site huts – location, ventilation, gas appliances
- Use of site main services
- Electricity – specialised equipment required?
- Fire fighting rules
- Waste disposal procedures
- Use of client's equipment
- Permit to work systems in force
- Hazardous substances in use on site by client
- Basic site arrangements, times, reporting, first-aid, fire
- Site boundaries and restricted areas

CONTROL OF CONTRACTORS ON SITE

The majority of the following measures are either legal requirements under the Management of Health and Safety at Work Regulations 1992, or will be required under the Construction (Design and Management) Regulations 1993 from 1st January 1994. They are essential for all contractor operations, however large or small the contract.

1. **Appointment/nomination of a person or team** to co-ordinate all aspects of the contract, including health and safety matters.

2. **A pre-contract commencement meeting** held with the contractor and sub-contractors as necessary, to review all safety aspects of the work. The contractor should also be asked to appoint a liaison person to ease later communication problems which may arise. Also communication paths should be developed to pass on all relevant safety information to those doing the work. Any permitted borrowing of equipment should be formally discussed at this time.

3. **Arrangement of regular progress meetings** between all parties, where health and safety is the first agenda item.

4. **Regular (at least weekly) inspections of the contractor's operations** by the client.

Management and Control of Contractors

5. **Participation in safety committees** on site by contractors should be a condition of the contract.

6. **Provision by the contractor of written method statements in advance** of undertaking particular work, as agreed. Work which this would apply to includes demolition, asbestos operations, work which involves disruption or alteration to main services or other facilities which cause interruption to the client's activities, erection of falsework or temporary support structures, and steel erection. An **essential** feature, but one often missing, is the stipulation that, in the event of the need for a deviation from the method statement, no further work will be done until agreement has been reached and recorded in writing between the client and the contractor on the method of work to be followed in the new circumstances.

7. The **formal reporting** to the client by the contractor of all lost-time accidents and dangerous occurrences, including those to sub-contractors.

8. The client must **set a good example** by the following of all site rules.

9. Provide adequate **safety propaganda** material, including posters and handbooks.

10. Allow no machinery on site until **documentation on statutory inspections** has been provided, including details of driver training and experience.

11. Monitor the contractor's **safety training** programme.

CONTRACT COMPLETION

The contractor should leave the worksite clean and tidy, removing all waste, materials, tools and equipment. This should be checked.

LEGAL REQUIREMENTS

Failure to manage contractors has wide implications under the Health and Safety at Work etc. Act 1974, where Sections 2, 3 and 4 can be applied to occupiers and contractors, depending upon the circumstances. Similarly, civil claims for damages can be made against occupiers as well as contractors. Provisions of the Management of Health and Safety at Work Regulations also apply, particularly in respect of the provision of information required to be in the possession of all employees.

SELF-ASSESSMENT QUESTIONS

1. Summarise the arrangements you make to select safe contractors in your workplace.

2. Develop a prequalification questionnaire suitable for use by contractors such as window cleaners, caterers and repair firms which, on completion, would help you evaluate their safety performance.

REVISION

Six parts to the control plan for contractors:

- Questionnaire to identify potentially safe contractors
- Hazard identification within the specification
- Check the bid and select contractor
- Put health and safety rules in the contract
- Control the contractor on site
- Check safe completion of work

Accident Investigation, Recording and Analysis

INTRODUCTION

"We .... are still unable to see a worker safely through a day's work. Why? Because we have not thoroughly analysed our accident causes. As a result of inadequate accident reporting we have had insufficient data to target in on unsafe tools, machines, equipment, or facilities. We also continue to place all of our safety bets on 'human perform-ance' to avoid hazards that are not being identified. We cannot expect to reduce our accident experi-ence by a solitary approach of attempting always to change human behaviour to cope with hazards. If there is a hole in the floor, we cannot reliably expect to avoid an accident by training all of the people to walk around the hole. It is far simpler to cover the hole." (David V. MacCollum, past President, American Society of Safety Engineers)

The hardest lessons to be learned in accident prevention come from the investigation of accidents and incidents which could have caused injury or loss. Facing up to those lessons can be traumatic for all concerned, which is one reason why investigations are often incomplete and simplistic. Nevertheless, the depth required of an investigation must be a function of the value it has for the organisation and other bodies which may make use of the results, such as enforcement agencies. Conducting one can be expensive in time. After the investigation, a standard system of recording and analysing the results should be used.

INVESTIGATION OF ACCIDENTS

Purpose

The number of purposes is large; the amount of detail necessary in the report depends upon the uses to be made of it. Enforcement agencies look for evidence of blame, claims specialists look for evidence of liability, trainers look for enough material for a case-study. From the viewpoint of prevention, the purpose of the investigation and report is to establish whether a recurrence can be prevented by the introduction of safeguards, procedures, training and information, or any combination of these.

The procedure

There should be a defined procedure for investigating all accidents, however serious or trivial they may appear to be. The presence of a form and checklist will help to concentrate attention on the important details. Supervisors of the workplace where the accident occurred will be involved; for less serious accidents they may be the only people who take part in the investigation and reporting procedure. Workers' representatives may also be involved as part of the investigating team.

The equipment

The following are considered as essential tools in the competent investigation of accidents and damage/loss incidents:

- Report form, possibly a checklist as a routine prompt for basic questions
- Notebook or pad of paper
- Tape recorder for on-site comments or to assist in interviews
- Camera – instant-picture cameras are useful (but further reproduction of them may be difficult, expensive or of poor quality)
- Measuring tape, which should be long enough and robust, like a surveyor's tape
- Special equipment in relation to the particular investigation, e.g. meters, plans

The investigation

Information obtained during investigations is given verbally, or provided in writing. Written documentation should be gathered to provide evidence of policy or practice followed in the workplace, and witnesses should be talked to as soon as possible after the accident. The injured person should also be seen promptly.

Key points to note about investigations are:

- Events and issues under examination should not be prejudged by the investigator
- Total reliance should not be placed on any one sole source of evidence
- The value of witness statements is proportional to the amount of time which passes between the events or circumstances described and the date of a statement or written record. (Theorising by witnesses increases as memory decreases)
- The first focus of the investigation should be on when, where, to whom and the outcome of the incident
- The second focus should be on how and why, giving the immediate cause of the injury or loss, and then the secondary or contributory causes
- The amount of detail required from the investigation will depend upon a) the severity of the outcome and b) the use to be made of the investigation and report

Accident Investigation, Recording and Analysis

- The report should be as short as possible, and as long as necessary for its purpose(s).

The report

For all purposes, the report which emerges from the investigation must provide answers to the following questions. Only the amount of detail provided should vary in response to the different needs of the recipients.

- What was the immediate cause of the accident/injury/loss?
- What were the contributory causes?
- What is the necessary corrective action?
- What system changes are either necessary or desirable to prevent a recurrence?
- What reviews are needed of policies and procedures (for example, risk assessments)?

It is not the task of the investigation report to allocate individual blame, although some discussion of this is almost inevitable. Reports are usually 'discoverable'; this means they can be used by the parties to an action for damages or criminal charges.

Whether the report is made on a standard form, or specially written, it should contain the following:

- A summary of what happened
- An introductory summary of events prior to the accident
- Information gained during investigation
- Details of witnesses
- Information about injury or loss sustained
- Conclusions
- Recommendations
- Supporting material (photographs, diagrams to clarify)
- The date, and be signed by the person or persons carrying out the investigation.

ACCIDENT RECORDING

Which injuries and incidents should be investigated and recorded? All which can give information useful to prevent a recurrence of the incident giving rise to loss or injury. Regulations may also define requirements, although reporting requirements are usually limited to the more serious injuries, and those incidents with the most serious consequences or potential consequences. Counting only these may mask the true extent of injuries and losses by ignoring the potential consequences of incidents which in fact have led to relatively trivial injuries or damage.

Standardised report forms kept at each workplace should be used, and returned to a central point for record-keeping and analysis. It is important that supervisory staff at the workplace carry out preliminary investigations and complete a report, as they should be accountable for work conditions and need to have personal involvement in failure (accidents and damage incidents). This demonstrates their commitment and removes any temptation to leave 'safety' to others who may be seen as more qualified.

ACCIDENT ANALYSIS

The incoming reports will need to be categorised and statistics collected so that meaningful information on causes and trends can be obtained. There are several ways of doing this, including sorting by the nature of the injury or body part involved; age group; trade; work location or work group; type of equipment involved. Selection of categories will depend upon the workplace hazards, but use of the categories and format of the RIDDO Regulations (See Module D) can be used, which will assist in making comparisons between works and national or industry group figures. One classification which will be found particularly helpful is breakdown by cause. The following gives heading examples; these can be further broken down if required:

- Falls   – from a height
  - on the same level, including trips and slipping
- Struck – by moving object
  - by vehicle
  - against fixed or stationary object
- Manual handling or moving of loads
- Machinery
- Contact with harmful substances
- Fire or explosion
- Electricity
- Other causes

Each of these categories, or, more usually, the aggregate numbers are converted into totals for a period, annually or at more frequent uniform intervals. Presentation in pictorial form is helpful for understanding; bar charts and pie charts are examples. Statistics used for comparison purposes are expressed in recognised ways using simple formulae to produce rates rather than raw numbers.

There are difficulties in using and comparing accident statistics, especially between different countries. This is because of under-reporting to the authorities collecting the data (which can be as high as 60%), differences

Accident Investigation, Recording and Analysis

between countries as to which accidents count as recordable (UK road traffic accidents do not count at present) and differences in the formulae used to calculate rates.

The UK **frequency rate** calculation uses a standard formula which copies neither of the above:

$$\frac{\text{Number of injuries} \times 100,000}{\text{Number of man-hours worked}}$$

The UK also uses **incidence rate** as an indicator, which is helpful where the number of man-hours worked is either low or not available:

$$\frac{\text{Number of injuries} \times 1000}{\text{Average number employed during the period}}$$

Confusingly, as there is no agreed or required formula,

the incidence rate most commonly used by the Government is:

$$\frac{\text{Number of fatal or major injuries} \times 100,000}{\text{Number at risk in the particular industry sector}}$$

LEGAL REQUIREMENTS

Employers and other 'responsible persons' with control over work premises as well as employees have a duty to report certain injuries suffered whilst at work, occupational diseases and defined dangerous occurrences. The RIDDO Regulations give detailed requirements (See Module D, Section 19). The employer is also required to keep an accident book on his premises, into which details of all injuries must be entered, as required by the Social Security Act 1975.

SELF-ASSESSMENT QUESTIONS

1. Find out the frequency rate for your workplace, compare it with the national average for your industrial classification. Can you calculate frequency rates for individual causes of injury? Is this knowledge useful?

2. Is 'carelessness' an adequate sole conclusion on an accident report?

REVISION

Eight features of the accident investigation process:

- Purpose
- Procedure
- Equipment
- Investigation

- Report
- Recording of results
- Analysis of results
- Presentation of results in meaningful format

Information Sources

INTRODUCTION

There is a great deal of information available on health and safety topics. The problem is that it is mostly unco-ordinated, in many places, and often written by specialists so that it cannot be easily understood by people who have to work with it. Information technology is moving towards the production of solutions to problems, by the combination of many of these sources.

Authoritative guides on particular topics are known as **primary sources**. The collected references to thee guides are **secondary sources**, and include bibliographies, reading lists, abstractions and indexes. Most information sources specialise as primary or secondary sources, with a small amount of overlap – some sources have both facilities. Technical articles in journals have reading lists, data sheets refer to larger data bases, and data bases often do not carry complete texts. Systems for information provision designed since about 1980 recognise the need for 'one-stop' information shopping for answers to problems, and avoid the temptation to cross-reference to a number of other sources.

Because of the numbers of sources, it is not possible to do more than provide a list of groups with some well-known examples, and a commentary on the material which is now available to carry the information.

PROVIDERS

Company safety policy	Organisation and arrangements
In-company safety services	Company safety staff, library
Corporate safety services	Central group resource, database
Enforcement agencies	HSE & local authorities (EHO) – advise and enforce
Government bodies and departments	Depts. of Trade and Employment, HMSO
Manufacturers	Product literature, updates, MSDS
Trade Associations	Handbooks, advice to members
Standards organisations	BSI, CEN, CENELEC, ISO
Subscription services	Magazines, journals, newsletters
Consultancies	External audits, information, advice
Voluntary safety bodies	RoSPA, British Safety Council
Professional bodies	IOSH, IChemE, BOHS
International safety bodies	ILO
Educational institutions	University programmes, colleges

MATERIAL

Most of our information is still provided on **paper**, and this is not likely to change despite introduction of new technologies. Some of them have problems associated – **photocopies** and **facsimile** transmission paper both suffer from ultraviolet light, so the image gradually disappears over time. Other photographic systems for holding information such as **microfiche** are more permanent, but can still be damaged and are inconvenient to read and copy, although they carry complete primary source material in a small space.

Computer files held at the workplace are increasingly used as information storage, although there is already evidence that new generations of computers are unable to decipher material stored on the older systems. **Computer networks**, set up between offices using the telephone network as a link, are able to share much information, which is often stored centrally on a mainframe. Access to such a system enables a large amount of information to be accessed at short notice at a local site. Disadvantages are cost, and the need to service the system and update regularly, which is also expensive in manpower.

The expense can be minimised by gaining access to someone else's information by using an electronic **data base**, and there are now a number of these which accept worldwide connections. Most are secondary sources, although the ability to carry full text and graphics is spreading. NIOSHTIC and HSELINE are the best known English-language data bases.

It is also possible to access versions of these and other databases at the workplace by installing a **CD-ROM** reader. Compact disc technology allows up to 280,000 pages of information to be stored on a single compact disc, which can be read by a personal computer. Recording onto CD-ROM is possible but requires expensive technology. Advantages of the system are cheapness, and the volume of material which can be stored and quickly accessed. Problem-solving systems are available through the Canadian Centre for Occupational Health and Safety (CCINFO) and Silver Platter (OSH-UK). Information can also be stored in learning programmes and combined with video into an **interactive system**, available on CD-ROM or on specially-adapted visual display screens.

SELF-ASSESSMENT QUESTIONS

1. Outline the steps you take at present to find health and safety information. How could these be improved?

2. Explain with examples the difference between a primary and secondary source of information. Mark the following sources as primary or secondary.

   - Your Company annual report
   - Full proceedings of an annual technical conference
   - Your shopping list
   - An international standard on eye protectors
   - A reference library

Communicating Safety

INTRODUCTION

Various forms of propaganda selling the 'health and safety message' have been used for many years. They are now widely felt to be of little value in measurable terms in changing behaviour and influencing attitudes to health and safety issues. Because of the long tradition of using safety propaganda as part of safety campaigns, however, there is a reluctance to abandon them. Possibly, this is because they are seen as constituting visible management concern whilst being cheap and causing minimal disturbance to production. This is in sharp contrast to the image of sales advertising and the degree of effort directed to this form of propaganda.

HOW CAN SAFETY MESSAGES BE EFFECTIVE?

- **Avoid negativity** Studies show that successful safety propaganda contains positive messages, not warnings of the unpleasant consequences of actions. Warnings may be ineffective because they fail to address the ways in which people make choices about their actions; these choices are often made subconsciously and are not necessarily 'rational' or logical. Studies also show that people make poor judgments about risks involved in activities, and are unwilling to accept a small loss of comfort or money as a trade against protection from a large but unquantified loss which may happen in the future.

"It won't happen here" – the non-relevance of the warning or negative propaganda needs to be combated. The short-term loss associated with some (but not all) safety precautions needs to be balanced by a positive short-term gain such as peace of mind, respect and peer admiration.

Safety propaganda can be seen as passing off the responsibility for safety from management to employees. Posters, banners and other visual aids used in isolation without the agreement and sanction provided by worker participation in safety campaigns can easily pass the wrong message. The hidden message can be perceived as "The management has done all it can or is willing to do. You know what the danger is, so it's up to you to be safe and don't blame us if you get hurt – we told you work is dangerous."

Management can have high expectations of the ability of safety posters to communicate the safety message. This will only be justified if they are used as part of a designed strategy for communicating positive messages.

- **Expose correctly** The safety message must be perceived by the target audience. In practice, this means the message must be addressed to the right people, be placed at or near to the point of danger, and have a captive audience.

- **Use attention-getter techniques carefully** Messages must seize the attention of the audience and pass their contents quickly. Propaganda exploiting this principle too readily can fail to give the message intended – sexual innuendo and horror are effective attention-getters, but as they may be more potent images they may only be remembered for their potency. Other members of the audience may reject the message precisely because of the use of what is perceived as a stereotype, for example 'flattering' sexual imagery can easily be rejected by parts of the audience as sexist and exploitative.

Strangely, the attention-grabbing image may be too powerful to be effective. An example to consider is the 'model girl' calendar. If the pictures are not regarded as sexist and rejected, they may be remembered. But who remembers the name of the sender? This is, after all, the point of the advertising.

- **Comprehension must be maximised** For the most effect, safety messages have to explain problems either pictorially or verbally in captions or slogans. To be readily understood, they must be simple and specific, as well as positive. Use of too many words or more than one message inhibits communication. Use of humour can be ineffective; the audience can reject the message given because only stupid people would act as shown and this would have no relevance to themselves or their work conditions.

- **Messages must be believable** The audience's ability to believe in the message itself and its relevance to them is important. Endorsement or approval of the message by peers or those admired such as the famous enhances acceptability. The 'belief factor' also depends upon the perceived credibility of those presenting the message. If the general perception of management is that health and safety has a low priority then safety messages are more likely to be dismissed because their contents and motivation are disbelieved.

- **Motivation and action must be achievable** Safety propaganda has been shown to be most effective when it calls for a positive action, which can be achieved without perceived cost to the audience and which offers a tangible and realistic gain. Not all of this may be possible for any particular piece of propaganda, and the major factor to consider is the positive action. Exhortation simply to 'be safe' is not a motivator.

DOES SAFETY PROPAGANDA WORK?

There is little evidence for the effectiveness of health and safety propaganda. This is mainly because of the difficulty of measuring changes in attitudes and behaviour which can be traced to the use of propaganda. For poster campaigns, experience suggests that any change in behaviour patterns will be temporary, followed by a gradual reversion to previous patterns, unless other activities such as changes in work patterns and environment are made in conjunction with the propaganda.

Limited observations by the authors show that the effectiveness of safety posters, judged by the ability of an audience to recall a positive safety message, is good in the short term only. One week after exposure to a poster, 90% of a sample audience could recall the poster's general details. 45% could recall the actual message on it. Two weeks after exposure, none of the audience remembered the message, and only 20% could recall the poster design. In both cases the attention of the audience was not drawn specifically to the poster when originally exposed to it (Unpublished research).

Safety propaganda can be useful in accident prevention, provided its use is carefully planned in relation to the audience, the message is positive and believable, and it is used in combination with other parts of a planned safety campaign. Safety posters which are not changed regularly become part of the scenery, and may even be counter-productive by giving a perceived bad image of management attitude.

REVISION

Six important elements in safety propaganda:

- Positive message
- Correct exposure
- Attention-getting
- Comprehension
- Belief
- Motivation

Techniques of Inspection

INTRODUCTION

Inspection for health and safety purposes often has a negative implication, associated with fault-finding. A positive approach based on **fact-finding** will produce better results, and co-operation from all those taking part in the process. Some experts believe that 'assurance' is a better description of the activity – there is a need to assure that the system is working properly (safely). To be effective, inspection of this type needs measurement of how good or bad things are, which can then be compared with standards set either locally, corporately or nationally. Corrective action can then be taken.

The objectives of inspections are :

- To identify hazardous conditions and start the corrective process
- To improve operations and conditions

There are a number of types of inspections:

- Statutory – for compliance with health and safety legislation
- External – by enforcement officials, insurers, consultants
- Executive – senior management tours
- Scheduled – planned at appropriate intervals, by supervisors
- Introductory – check on new or reconditioned equipment
- Continuous – by employees, supervisors, which can be formal and preplanned, or informal.

For any inspection, knowledge of the plant or facility is required, also knowledge of applicable Regulations, standards and codes of practice. Some system must be followed to ensure that all relevant matters have been considered, and an adequate reporting system must be in place so that remedial actions necessary can be taken and that the results of the inspection are available to management.

PRINCIPLES OF INSPECTION

Before any inspection, certain basic decisions must have been taken about aspects of it, and the quality of the decisions will be a major influence on the quality of the inspection and on whether it achieves its objectives. The decisions are reached by answering the following questions:

1. **What needs inspection?** Some form of checklist, specially developed for the inspection, will be helpful. This reminds those carrying out the inspection of important items to check, and serves as a record. By including space for 'action by' dates, comments and signatures, the checklist can serve as a permanent record.

2. **What aspect of the items listed needs checking?** Parts likely to be hazards when unsafe – because of stress, wear, impact, vibration, heat, corrosion, chemical reaction or misuse – are all candidates for inspection, regardless of the nature of the plant, equipment or workplace.

3. **What conditions need inspection?** These should be specified, preferably on the checklist. If there is no standard set for adequacy, then descriptive words give clues to what to look for – items which are exposed, broken, jagged, frayed, leaking, rusted, corroded, missing, loose, slipping, vibrating.

4. **How often should the inspection be carried out?** In absence of statutory requirements, or guidance from standards and codes of practice, this will depend upon the potential severity of the failure if the item fails in some way, and the potential for injury. It also depends upon how quickly the item can become unsafe. A history of failures and their results may give assistance.

5. **Who carries out the inspection?** Every **worker** has a responsibility to carry out informal inspections of his or her part of the workplace. **Supervisors** should plan general inspections, and take part in periodic inspections of aspects of the workplace considered significant under the above guideline. **Workers' representatives** may also have rights of inspection, and their presence should be encouraged where possible. **Management** inspections should be made periodically; the formal compliance inspections should take place in their presence.

TECHNIQUES OF INSPECTION

The following observations have been of assistance in improving inspection skills:

1. Those carrying out inspections must be properly equipped to do so, having necessary knowledge and experience, and knowledge of acceptable performance standards and statutory requirements. They must also comply fully with local site rules, including the wearing or use of personal protective equipment, as appropriate, so as to set an example.

2. Develop and use checklists, as above. They serve to focus attention and record results, but must be relevant to the inspection.

3. The memory should not be relied upon. Interruptions will occur, and memory will fade, so notes must be taken and entered onto the checklist, even if a formal report is to be prepared later.

4. It is desirable to read the previous findings before starting a new inspection. This will enable checks to be made to ensure that previous comments have been actioned as required.

5. Questions should be asked, and the inspection should not rely upon visual information only. The "what if?" question is the hardest to answer. Workers are often undervalued as a source of information about actual operating procedures and of opinions about possible corrections. Also, systems and procedures are difficult to inspect visually, and their inspection depends upon the asking of the right questions of those involved.

6. Items found to be missing or defective should be followed up and questioned about, not merely recorded on the form. Otherwise, there is a danger of inspecting a series of symptoms of a problem without ever querying the nature of the underlying disease.

7. All dangerous situations encountered should be corrected immediately without waiting for the written report, if their existence constitutes a serious risk of personal injury or significant damage to plant and equipment.

8. Where appropriate, measurements should be taken of conditions. These will serve as baselines for subsequent inspections. What cannot be measured cannot be managed.

9. Any unsafe behaviour seen during the inspection should be noted and corrected, such as removal of machine guards, failure to use personal protective equipment as required, or smoking in unauthorised areas.

10. Risk assessments should be checked as part of the inspection process.

SELF-ASSESSMENT QUESTIONS

1. On the next two pages, you will find headings for a safety inspection checklist, with more detailed subheadings. Tick the boxes in pencil to indicate the kinds of inspections now made in your workplace for each category and the standards which apply, putting a cross where none is made. Then repeat, using a pen, indicating what the ideal inspections would be for each.

2. On the third page, you will find a similar set of headings, to complete those previously given. This time, the subheadings are missing and are for you to fill in as appropriate to your workplace. Then, indicate what the ideal inspections would be.

REVISION

Two objectives of inspections:

- Identification of hazards
- Improvement of operations/conditions

Six types of inspection:

- Statutory
- External
- Executive
- Scheduled
- Introductory
- Continuous

Ten techniques of value:

- Have necessary experience and knowledge
- Use checklists
- Write things down
- Read previous reports first
- Ask questions
- Follow up on problems
- Correct dangerous conditions at once
- Measure and record where possible
- Correct unsafe behaviour seen
- Check risk assessments

Techniques of Inspection

Part One	Inspection before each use	Weekly	Monthly	Annually	Company/Group Standard	National Standard	International Standard	No current standards
Environmental Factors								
Lighting								
Dusts								
Gases								
Fumes								
Noise								
Hazardous Substances								
Flammables								
Acids & bases								
Toxics								
Carcinogens								
Production Equipment								
Machinery								
Pipework								
Conveyors								
Power Generation Equipment								
Steam equipment								
Gas equipment								
Generators								
Electrical Equipment								
Cables, circuits								
Switches, sockets								
Extensions								
Personal Protective Equipment								
Eye protection								
Clothing								
Helmets/head protection								
RPE								
Footwear								
Hand protection								
First Aid and Welfare								
Washing facilities								
Showers								
Toilets								
First-aid facilities								

Part One	Inspection before each use	Weekly	Monthly	Annually	Company/Group Standard	National Standard	International Standard	No current standards
Fire Protection Equipment								
Hoses and extinguishers								
Alarm systems								
Sprinklers								
Access Routes, Roadways								
Roads								
Pavements								
Crossing points								
Signs								
Lifts								
Stairways								
Emergency routes								
Access Equipment								
Ladders								
Stepladders								
Trestles								
Cradles								
Materials Handling Equipment								
Chains								
Ropes								
Forklifts								
Specialised equipment								
Transport								
Cars								
Internal vehicles								
Road load vehicles								
Containers								
Disposal								
Storage								
Cylinders								

Module C Section 12

Techniques of Inspection

Part Two	Inspection before each use	Weekly	Monthly	Annually	Company/Group Standard	National Standard	International Standard	No current standards
Hand tools								
Storage facilities								
Structures								
Structural openings								
Warning and signalling equipment								
Miscellaneous								

FOR ALL SECTIONS

Health and Safety at Work etc Act 1974
Management of Health and Safety at Work Regulations 1992
Workplace (Health, Safety and Welfare) Regulations 1992
Provision and Use of Work Equipment Regulations 1992
Manual Handling Operations Regulations 1992
Personal Protective Equipment at Work Regulations 1992
Factories Act 1961
Offices, Shops and Railway Premises Act 1963
Fire Precautions Act 1971

Guidance (HSE/C)

Associated Approved Codes of Practice and Guidance on above Regulations
HS(G)65   Successful health and safety management

THE SAFETY POLICY

Guidance (HSC/E)

Guidance on the implementation of safety policies for the construction industry – IAC/L1

Effective policies for safety and health (Accident Prevention Advisory Unit)

Writing your health and safety policy statement: guide to preparing a Safety policy statement for a small business

SAFE SYSTEMS OF WORK

Guidance (HSE)

IND(G)76L Safe systems of work (leaflet)

TRAINING

Abrasive Wheels Regulations 1970
Woodworking Machines Regulations 1974
Health and Safety (First-aid) Regulations 1981
Health and Safety (Training for Employment) Regulations 1990

Guidance (HSC/E)

COP 20	Standards of training in gas safety installations
COP 26	Rider operated lift trucks – operator training
COP 42	First-aid at work
GS28/4	The safe erection of structures: Legislation/training
GS39	Training of crane drivers/slingers
HS(R)9	A Guide to the Woodworking Machines Regulations 1974

MAINTENANCE

Electricity at Work Regulations 1989
Control of Substances Hazardous to Health Regulations 1988

British Standards

BS4430	Recommendations for industrial trucks Part 2: operation and maintenance
BS5405	Code of practice for maintenance of electrical switchgear (See also BS6423 and BS6626)
BS5720	Code for mechanical ventilation and air conditioning in buildings
BS6071	Specification for periodic inspection and maintenance of transportable gas containers for dissolved acetylene
BS6913	Operation and maintenance of earth moving machinery

Guidance (HSC/E)

PM26	Safety at lift landings
PM45	Escalators; periodic thorough examination
EH48	Legionnaires Disease
HS(G)54	The maintenance, examination and testing of local exhaust ventilation

Deadly maintenance – Plant & machinery: a study of fatal accidents at work
Roofs: a study of fatal accidents at work
A study of fatal accidents at work

CONTRACTORS

The Health and Safety Information for Employees Regulations 1989
Occupiers Liability Act 1957 and 1984

Module C Section 13

Selected References

Guidance (HSE)

Managing Health and Safety in Construction (CON-IAC)

ACCIDENT INVESTIGATION

Reporting of Injuries, Diseases and Dangerous Occurrences Regulations 1985 (RIDDOR)
Social Security Act 1975 (as amended)
Social Security (Industrial Injuries & Adjudication)(Miscellaneous Amendments) Regulations 1986
Social Security (Industrial Injuries)(Prescribed Diseases) Regulations 1985

Guidance (HSC/E)

HSE 119	Reporting an injury or dangerous occurrence
HSE 17Z	Reporting a case of disease
HS(R)23	A Guide to RIDDOR

INFORMATION

PANTRY, S. Health and Safety: a guide to sources of information. CPI, The Grey House, Broad Street, Stamford, Lincs PE9 1PR. ISBN 0 90 6011 29 9 (£10.00)

TECHNIQUES OF INSPECTION

Construction (Lifting Operations) Regulations 1961
Construction (Working Places) Regulations 1966
Offices, Shops and Railway Premises (Hoists and Lifts) Regulations 1968
Power Presses Regulations 1965
Pressure Systems and Transportable Gas Containers Regulations 1989

Guidance (HSC/E)

COP 37	Safety of pressure systems
COP 38	Safety of transportable gas containers
PM27	Construction hoists

Notes

Module D:
Law

Introduction to The English Law Module

Much of this Module covers the legal framework and requirement in England and Wales. There is no difference between the legal systems of these countries. The system in Scotland and Northern Ireland is different, although not such as to make significant differences in the standards of health and safety required to be observed. Differences in the organisation and administration of the law, in procedures and terminology, will, however, be significant for the health and safety practitioners resident in those countries.

On matters of health and safety law, the Scottish system is essentially the same as the English system. Acts of Parliament state whether or not they apply in Scotland – Section 182 of the Factories Act 1961 states that the Act applies in Scotland. Where a statute is not applicable, legislative harmonisation is necessary.

Northern Ireland also has its own legal system, governed by the United Kingdom Parliament. In most circumstances additional legislation is required by Acts, Regulations or Orders to introduce health and safety provisions into Northern Ireland. Until recently there has been a lag of about two years between introduction of legislation in England and the appearance of its counterpart in Northern Ireland, but this should change after the introduction of the European Single Market.

The selection of statutes to be discussed in this Module was made on the basis that some affect everyone, and should be included, and there should also be discussion of legislation which is either important or can serve as an example of a type.

Some of the Regulations discussed include technical terms either in the Regulations themselves or in the accompanying Code of Practice. These terms have been defined in other Modules, and are not defined in the following Sections.

The reader will probably be aware that Regulations are being made under the Health and Safety at Work etc. Act 1974 to implement European Community (EC) Directives. These require similar laws to be passed in each member State of the Community on their subjects, which must at least meet the requirements of the Directive concerned and may exceed them. In general, UK laws made in response to Directives on health and safety at work exceed the Directives because they apply to the self-employed as well as employees, employers and other duty-holders. As a member State, the UK is required to activate laws implementing Directives by no later than the deadline given in the Directive concerned. For the six sets of health and safety Regulations described in later Sections of this Module which were laid before Parliament in 1992, the activation date was January 1st 1993, to coincide with the beginning of the Single Market in Europe.

This book does not cover the Regulations made to comply with Directives on Product and Machinery Safety, which impose standards on products and equipment rather than on people and systems. There are considerable links between the two types of Regulations, which are discussed briefly in this Module. For a detailed treatment, the reader is directed to the various Approved Codes of Practice and Guidance published with the Regulations.

Self-assessment questions have not been provided for some of the Sections in this Module. Decisions about the need for questions were taken on the basis of the need to self-test understanding of basic and new requirements.

The English Legal System

INTRODUCTION

The English legal system uses different types of courts to hear different types of cases. Some hear only criminal matters, where guilt is the matter at issue, and others hear only civil cases, where the aim (for health and safety purposes) is to obtain compensation for a person suffering injury or damage. It should be said that there is a financial limit on much of the jurisdiction of the civil courts, which is otherwise wide-ranging. There are numerous remedies available in civil courts, of which damages are but one. In some instances it is possible for courts to hear both types of case. The senior courts hear the more important cases and appeals from the lower courts. Appeals can generally only be made by the defendant, although points of law can be raised as grounds for appeal by either side. The prosecution cannot appeal against a sentence in the Magistrates' Court, but there are now limited rights of appeal for it in the Crown Court. In civil cases, the right of appeal and the grounds upon which appeal may lie are very much wider.

The framework of the court system is shown in the accompanying diagram.

MAGISTRATES' COURTS

Magistrates' Courts hear both civil and criminal cases, and most criminal prosecutions begin in these courts (the majority also ending there as well). The civil jurisdiction of Magistrates' Courts extends to the recovery of certain debts, such as income tax, electricity, gas and water charges. Care proceedings are dealt with in the Family Proceedings Court. Magistrates' Courts also deal with cases brought by the enforcing authorities under health and safety legislation, although as the court's powers are relatively limited these are sometimes committed to the Crown Court for trial and/or sentence.

Cases are heard by members of the public who are appointed as Magistrates (Justices of the Peace). They need have no legal qualifications, acting only as decision-makers on both fact and law, and take legal advice from the Clerk to the Justices for each Bench. Each "commission area" – usually a county in the Shires – is divided into petty sessional divisions. A Bench (the Magistrates) serves the petty sessional division. Each individual court normally sits with three Magistrates, and a qualified Court Clerk. Magistrates are unpaid and carry out their duties in addition to their normal employment. There are some professional legally-qualified Magis-

trates, however, known as "stipendiaries", who are not required to sit with other Magistrates and can adjudicate by themselves.

The powers of Magistrates Courts are limited to fines of up to £5,000 and/or up to 6 months imprisonment. For breaches of the 'general duties' Sections of the Health and Safety at Work etc. Act 1974 (Sections 2 – 6), maximum fines have been increased to £20,000. Fines at this level can also be imposed upon those ignoring Prohibition Notices.

Some matters are triable **summarily** (only at the Magistrates' Courts), some are **triable either way** (the defendant and the prosecution can make representations about whether they think a jury trial and higher sentencing powers are appropriate, and the Bench decides whether it is prepared to hear the case in their court). The defendant is then given an opportunity to select the court of his or her choice, having been told that the Magistrates may still decide to send the case to the higher court for sentence if they feel their own powers of sentencing are insufficient.

Some offences are triable only on **indictment** (murder, for example) at the Crown Court. For these most serious cases, the functions of the Examining Magistrates are set out in Sections 4-8 of the Magistrates' Courts Act 1980. In short, the function is to hold committal proceedings in order to enquire into the evidence of the prosecution, and to examine the prosecution case to decide whether there is sufficient evidence to put the defendant on trial by jury. The test is whether the prosecution has brought forward sufficient evidence to satisfy the Examining Magistrates that there is a triable issue to be put before the jury.

TRIBUNALS

Tribunals are creatures of statute – and provide a complex system with different tribunals being regulated by different statutory provisions. They are not courts – they are bodies established to take decisions in particular areas of the law, and they are regulated by the legal instrument which brings them into being. Examples of tribunals are those covering employment disputes including some health and safety matters (industrial tribunals), and rent disputes (rent tribunals). Tribunal members are appointed, and are not necessarily legally qualified.

The growth and success of the tribunal system may be due to the fact that they provide a cheap and less formal way of resolving disputes. Activities of tribunals are controlled ultimately by the conventional court system

– the Divisional Court of the Queens Bench Division hears appeals which have exhausted the full tribunal mechanism. Appeals against tribunal decisions are allowed on questions of law; decisions will be referred back if the tribunal has acted improperly, exceeded its jurisdiction or refused to hear a case.

COUNTY COURTS

The County Court is the junior of the two main civil courts, the other being the High Court. It only hears civil cases, which range from landlord/tenant disputes to hire purchase repossessions. All cases in this court are heard by a judge or a registrar (the junior judge), the latter normally hearing the less important cases. Its jurisdiction is extensive, and this brief summary cannot hope to do more than indicate its range.

CROWN COURTS

The Crown Court is responsible for hearing the more serious criminal cases, particularly those not tried, and those which cannot be tried (indictable offences) by the Magistrates' Court. It also deals with cases there the defendant has elected trial by jury (if the offence is "triable either way" – in either court), or because the Magistrates have determined that a particular offence should be tried by a jury.

Crown Courts also hear appeals against convictions by Magistrates Courts, as well as other matters. Whilst no jury is present, lay Magistrates are.

Crown Courts are widespread geographically, and are graded according to the seriousness of the offence which in turn determines the seniority of the judge who will hear the case. The Old Bailey is the most famous and senior of the Crown Courts. Appeals from these Courts are heard by the Court of Appeal (Criminal Division).

THE HIGH COURT

All the important civil cases are heard in the High Court, which also has some jurisdiction over criminal appeals. Its work is divided into three main divisions:

The Queens Bench Division administers primarily the common law. It hears cases dealing with contracts, or with civil wrongs (known as **torts**). The tort of negligence, for example, is the means by which compensation may be obtained following an accident at work. The division also hears those cases which cannot be brought to the County Court – such as those where claims involve more than £5,000. Within the Division there are two specialist Courts, Commercial (hearing major commercial disputes in private where the judge acts as arbitrator) and the Admiralty Court (hearing maritime disputes).

The Chancery Division administers primarily equity. It specialises in hearing cases involving money, such as tax, trusts, bankruptcy, disputes over wills, or with the Inland Revenue.

The Family Division deals with family disputes which cannot be handled by the Magistrates Courts. These include defended divorces, separation, child custody, adoption and distribution of family assets on marriage breakdown.

Within each High Court Division there is a Divisional Court, presided over by three judges at each sitting, as opposed to the single judge in the High Court.

The Divisional Court of the Queens Bench Division hears appeals from Magistrates' Courts, or from the Crown Court on appeal from a Magistrates' Court on criminal matters, and can undertake judicial reviews. It also overseas the activities of the junior courts and the tribunals. Further appeal lies from the Divisional Court to the House of Lords.

The Divisional Court of the Chancery Division deals with appeals from County Courts on bankruptcy matters.

The Divisional Court of the Family Division hears appeals from Magistrates Courts in domestic and matrimonial cases.

THE COURT OF APPEAL

This Court is divided into two distinct courts, one for hearing appeals arising from criminal proceedings and one for hearing criminal appeals.

The Civil Division of the Court of Appeal hears appeals from County Court and High Court decisions. The Court rehears the case, listening only to legal argument from counsel and relying on transcripts taken from the previous hearings. Most appeals are based on points of law, but some are based on the claim that the original judge may have drawn wrong conclusions from the evidence. Three Lord Justices of Appeal constitute the Court, and the decisions they reach can be made on a majority vote. The Master of the Rolls is the most senior of the Lords Justices of Appeal.

The Criminal Division of the Court of Appeal hears appeals from the Crown Court. This Court can overturn a conviction, vary the sentence imposed or substitute a conviction for another offence. Appeals can be made by the defendant, arising out of points of law. Also, where a defendant is acquitted, the Attorney General may refer a point of law involved in the case to the Court of Appeal in order to have a ruling for future cases. There is now an important provision enabling the prosecution to refer to the Court of Appeal a sentence imposed in the Crown Court which it regards as being unduly lenient. Generally, appeals in this Division are heard by two judges; if they cannot agree the appeal is reheard in front of three. The Lord Chief Justice is the head of the Division.

Appeals from the decisions of the two Divisions of the Court of Appeal are made to the House of Lords.

THE HOUSE OF LORDS

This is the highest appeal body for both criminal and civil cases, short of the European Court. All cases heard at this level require approval from the Court of Appeal, or from the Appeals Committee of the House of Lords. Permission is only given if the appeal is of significant legal importance, or in the wide public interest (such as the "Spycatcher affair"). In exceptional cases, an appeal can "leap-frog" the Court of Appeal to be heard immediately by the House of Lords. For this to happen, the High Court judge and the Lords must certify that there is a legal point of public importance involved which may alter previous legal decisions.

Judges in this Court are known as the Law Lords, and all are professional judges who have been awarded a life peerage. Also, the Lord Chancellor and peers who have held high judicial office can sit as judges in the Court. Technically, any Lord can be involved, but in practice this never happens. A minimum of three is required to make a decision, in practice there are always five. Due to the high cost involved in bringing cases to the Lords, very few criminal cases reach the Court. Most cases involve tax, commercial and property disputes.

THE JUDICIAL COMMITTEE OF THE PRIVY COUNCIL

Although this Committee does not form a part of the English legal system, it exerts a considerable influence over it. It is the highest appeal forum for cases within the British Empire. Obviously, as the Empire has declined, its activities have been restricted to the Commonwealth countries, but it also hears cases from the Isle of Man and the Channel Islands. Decisions of the Judicial Committee are respected by countries which have based their legal system on English law. Essentially, judges involved in the activities of the Committee are those who sit in the House of Lords, but there can also be an involvement of Commonwealth judges.

EUROPEAN COURTS

The European Court of Justice, sitting in Luxembourg, adjudicates upon the law of the European Economic Community, and its decisions are binding on British Courts by reason of the European Communities Act 1972. It gives rulings regarding the interpretation and application of provisions of European Community law referred to it by Member States. A single judgment is given, from which there is no right of appeal. Any decisions it makes are enforceable throughout the network of courts and tribunals within each of the Member States.

The European Court of Human Rights enforces the agreed Community standards on the protection of human rights and fundamental freedoms.

EUROPE'S ROLE IN ENGLISH HEALTH AND SAFETY LAW

Some current health and safety legislation, and much of what is proposed for the future, originates within the European Community (EC). Administration of the EC is carried out through a network of four bodies:

The Commission, led by a Board of Commissioners of representatives from each Member State. The Commission administers the law throughout the EC, much like the civil service does in this country. Proposals for future legislation originate from the Commission in the form of Directives, sent for approval to the Council of Ministers.

The Council of Ministers is composed of members of government from each Member State, and is responsible for decision-making within the EC – it is similar to the Government in this country.

The European Parliament is formed of representatives elected from constituencies from within Member States, and consults on and debates all proposed legislation.

The European Courts – see above.

Apart from the general provisions of the Treaty of Rome, which is signed by all Member States, there are three elements or instruments of European law:

European Regulations passed by the Council of Ministers, which are adopted immediately into the legal framework of each Member State, unlike

Directives, which set out objective standards to be achieved but allow Member States to decide on how their own individual legal frameworks will be altered to adopt them. In the UK, the most common method of implementation of Directives is by Statutory Instrument.

Decisions are reached in cases which are taken to the European Courts, and affect only the company or individual concerned.

SELF-ASSESSMENT QUESTIONS

1. Write notes on the structure and functions of Magistrates' Courts, and explain the extent of the powers of the Court in relation to health and safety legislation.

2. How do European Directives affect health and safety in the UK?

REVISION

The English legal system is summarised in the diagram below.

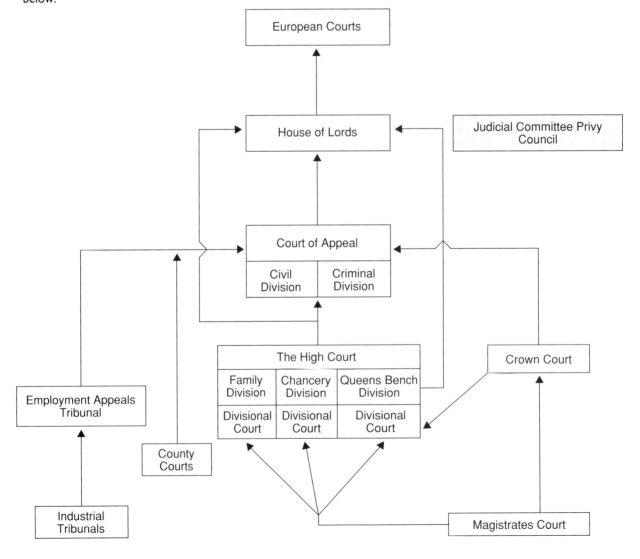

Statute and Common Law

INTRODUCTION

"The law" concerning health and safety matters is a mixture of criminal law set out in statutes, and known as statute law, and common law, which provides a means of compensation for injuries or damage suffered because of failure of another party to comply with statute law or to carry out any of the duties established by common law over the years. Breaches of statute law are criminal offences for which financial penalties can be imposed, and even prison sentences in special and relatively uncommon situations. Individual cases are always judged on their particular facts. "The law" stands as interpreted by the courts and announced in court decisions. Decisions of higher courts are binding on those below – these are called **precedents**. Lower court decisions may be helpful in reaching decisions; these are called **persuasive**. Decisions of the House of Lords bind all lower courts, and can only be overridden by the House itself changing its mind, or by an Act of Parliament.

STATUTE LAW

Statute law is the written law of the land and consists of **Acts of Parliament**, and the Rules, Regulations or Orders made within the parameters of the Acts. Acts of Parliament usually set out a framework of principles or objectives, and use specific Regulations or Orders to achieve these, sometimes after a delay written in to the Act. Most of the laws relating to health and safety at work are contained within a large body of Regulations (some Rules and Orders are still operative) aimed at specific requirements and enforced by the Health and Safety Executive or Local Authorities with delegated powers to do so.

The Act of Parliament is therefore the primary or principal legislation; Regulations made under Acts are secondary, subordinate, delegated legislation. For example, the Health and Safety at Work etc. Act 1974 is a primary Act of Parliament, and Regulations made under it have the same terms of reference and applicability – the Noise at Work Regulations 1989 and the Electricity at Work Regulations 1989 being only two of many.

Regulations are made by Ministers of the Crown where they are enabled to do so by an Act of Parliament. Regulations have to be laid before Parliament, but most do not need a vote and become law on the date specified in the Regulation. They can still be vetoed by a vote within forty days of being laid before Parliament. Regulations can apply to employment conditions generally, for example the Control of Noise at Work Regulations; they can control specific hazards in specific industries (the four Construction Regulations); and they can have the most general application (Reporting of Injuries, Diseases and Dangerous Occurrences Regulations). Section 15 of the Health and Safety at Work Act gives a lengthy list of the purposes for which Regulations can be made under it.

Health and safety Regulations are made by the Secretary of State (usually for Employment), acting on proposals made by the Health and Safety Commission (HSC) or after consultation with the Commission. Normally, the HSC issues Consultative Documents to all interested parties before making such recommendations.

Orders and **Rules** technically give the force of law to an executive action. Thus, Commencement Orders can be made by the Secretary of State which activate parts of Acts or Regulations.

Approved Codes of Practice can supplement Acts and Regulations, in order to give guidance on the general requirements which may be set out in the legislation (thus effectively enabling the legislation to be kept up to date by revising the Code of Practice rather than the law). The HSC has power to approve Codes of Practice of its own, or of others such as the British Standards Institution. Approval cannot be given without consultation, and consent of the Secretary of State. Failure to comply with an Approved Code of Practice is not an offence of itself, but failure is held to be proof of contravention of a requirement to which a Code applies unless a defendant can show that compliance was achieved in some equally good way (Section 17, Health and Safety at Work Act).

Guidance Notes are documents issued by the HSC or the Health and Safety Executive (HSE) as opinions on good practice. They have no legal force, but because of their origin and the experience employed in their production, they will be "persuasive" in practice to the lower courts and useful in civil cases to establish reasonable standards prevailing in an industry. Approved Codes and Guidance Notes can therefore be said to have a "quasi-legal" status, rather like the Highway Code.

Breaches of statute law can be used in civil claims to establish negligence, unless this is prohibited by the Act or Regulation itself. The Health and Safety at Work Act contains such a provision, so breaches of it cannot be used to support civil claims for compensation.

TYPES OF STATUTORY DUTY

There are three types of duty imposed by statute, which allow different responses to hazards. The types are: absolute duty, duty to do what is practicable, and duty to take steps that are reasonably practicable. Over the years, a body of case law has built up which gives guidance on the meaning of these duties in practice. Mostly, the cases have been decided in common law, in the course of actions for personal injuries which were based on alleged breach of statutory duty.

1. **Absolute duty**. (This is also known as **strict liability**). There are circumstances when the risk of injury is so high unless certain steps are taken, and in consequence Acts and Regulations have recognised these by placing an absolute duty on the employer to take specific steps to control the hazard. The words "must" or "shall" appear in the Section or Regulation to indicate this – there is no choice or evaluation of risk or feasibility to be made by the employer. An example occurs in (the unamended) Section 14 of the Factories Act 1961, which requires "the fencing of every moving part of every prime mover". The absolute nature of the duty has been upheld by courts, even in circumstances where a machine has become practically or financially unusable because of the guarding requirement. However, an "easement" is sometimes given where such absolute requirements are subject to a "defence" clause where, in any proceedings for an offence consisting of a contravention of an absolute requirement, it shall be a defence for any person to prove that he took all reasonable steps and exercised all due diligence to avoid the commission of that offence. This is known as the "due diligence" defence, and an example can be found in Regulation 29 of the Electricity at Work Regulations 1989.

2. **"Practicable"**. Some Regulations specify that steps must be taken "so far as is practicable", for example (the unamended) Regulation 10 of the Woodworking Machines Regulations 1974 which requires sufficient clear and unobstructed space around every woodworking machine to be maintained "so far as is ...... practicable" to enable the work at the machine to be done without risk to employees. "Practicable" means something less than physically possible. To decide on whether the requirement can be achieved the employer would have to consider current technological knowledge and feasibility, not just difficulty of the task, its inconvenience or cost. Some degree of reason can be applied, and current practice.

3. **"Reasonably practicable"**. The phrase "so far as is reasonably practicable" qualifies almost all the general duties imposed on employers by the Health and Safety at Work Act. Its use allows the employer to balance the cost of taking action (in terms of time and inconvenience as well as money) against the risk being considered. If the risk is insignificant against the cost, then the steps need

not be taken. This analysis of cost versus benefit must be done prior to any complaint by an enforcing authority. An employer would need to keep up to date with health and safety developments affecting his activities, as steps might become reasonably practicable over a time. Also, a single analysis of precautions could be insufficient in the sense that other less effective but cheaper alternatives might be available in addition to the one being evaluated. "Reasonably practicable" steps are those which are the best under the circumstances.

COMMON LAW

Common law has evolved over hundreds of years as a result of the decisions of courts and judges. Common principles or accepted standards fill the gaps where statute law has not supplied specific requirements. The accumulation of common law cases has resulted in a system of precedents, or decisions in previous cases, which are binding on future similar cases unless over-ruled by a higher court or by statute.

Case law on health and safety did not appear until the 19th. century, when the rapid introduction of industrial technology led to an equally rapid rise in work injuries. In 1840 a young female mill worker successfully sued her employer and was awarded damages [Cottrell v Stocks (1840) Liverpool Assizes]. Claims at common law were often resisted successfully by employers, especially if there had been any contribution to the accident by the worker's own actions or if the worker had claimed additionally for compensation under the Workman's Compensation Act. All these restrictions on claims were removed by reforms (but not until after the Second World War), which were completed by the Law Reform (Personal Injuries) Act 1948.

As a result of the cases which have been decided over the years, there is now a body of precedents which defines the duties of both employers and employees at common law. The employer must take reasonable care to protect his employees from the risk of foreseeable injury, disease or death at work by the provision and maintenance of a safe place of work, a safe system of work, safe plant and equipment, and reasonably competent fellow employees. The basis of the employer's duty to employees derives from the existence of a contract of employment, but the duty to take reasonable care may also extend to matters which affect others but are within the employer's control.

Statute and Common Law

To avoid common law claims (and to avoid the accidents which prompt them) the employer has to do what is reasonable under the particular circumstances. This will include providing a safe system of work, instruction on necessary precautions and indicating the risks if the precautions are not followed.

The employer can also be held liable for the actions of his employees causing injury, death or damage to others, providing the actions were committed in the course of employment. This liability of the employer is known as **vicarious liability**.

Employees have a general duty of care in common law towards themselves and to other people – so they can also be sued by anyone injured as a result of lack of reasonable care on their part. In practice this happens rarely, and the liability is assumed by the employer under vicarious liability.

The burden of proof in common law cases rests with the plaintiff – the injured party. He must show that he was owed a duty by whoever he is suing, that there was a breach of the duty, and that as a result of that breach he suffered damage. The plaintiff can be helped by pro-cedural rules and rules of evidence, one of which saves him having to show exactly how an accident happened if all the circumstances show that there would not have been an accident except because of lack of reasonable care on the part of the person being sued. This is known as **res ipsa loquitur** – "let the facts speak for themselves".

Defences can be raised against common law claims: there was no negligence, there was no duty owed, the accident was the sole fault of the employee, the accident did not result from the lack of care, contributory negligence (very rarely a complete defence), and **volenti non fit injuria** – the employee knowingly accepted the risk.

Civil actions have to be started within three years of the date when the accident happened, or the date when the injured person found there was an injury which was the fault of the employer. The court has discretion to extend the three-year period in appropriate cases.

Damages awarded are difficult to quantify, but they depend upon loss of a faculty, permanent nature of the injury and its effect on the ability to earn a living, the expenses incurred and a range of other factors. There is a "tariff" of awards used informally by judges, but this is often not followed. English courts do not use the jury system (except in libel cases), which awards damages in other jurisdictions. Awards may be reduced by a percentage if there is found to be a contribution to the injury caused by the plaintiff's own negligence, but the trend of the courts is to minimise this as a significant factor.

Self-Assessment Questions

1. Outline the main differences between common law and statute law.

2. Explain the meaning of "reasonably practicable".

Revision

Two types of applicable law:

- Statute law
- Common law

Statute law:

Made by Acts, Regulations, Orders, Rules, supplemented by Approved Codes of Practice, Guidance Notes

Three types of statutory duty:

- Absolute (strict liability)
- Practicable
- Reasonably practicable

Common law:

- Based on judicial precedent
- Involves a general duty of care
- Breach of statutory duty can usually establish breach of general duty
- Defences can be based on:
    Lack of a duty
    Lack of negligence
    Sole fault of injured person
    Contributory negligence
    Voluntary acceptance of the risk

The Health and Safety at Work Etc. Act 1974

INTRODUCTION

This Act is the major piece of health and safety legislation in Great Britain. It provides the legal framework to promote, stimulate and encourage high standards. Previous Acts had concentrated upon prescription of solutions within the law; a series of Factories Acts in particular had raised standards progressively since 1833 by stipulating what measures needed to be taken in factories, and although the legislative base was gradually widened to include other kinds of premises, modern work practices and businesses were no longer fully covered by legislation by the third quarter of the century. Consultations carried out by the Robens Committee between 1970 and 1972 produced the basis of a new type of law – one which placed responsibility on employers and employees together to produce their own solutions to health and safety problems, subject to the test of reasonable practicability (see Section 3).

The Act introduced for the first time a comprehensive and integrated system dealing with workplace health and safety and the protection of the public from work activities. By placing duties of a general character upon employers, employees, the self-employed, manufacturers, designers and importers of work equipment and materials, the protection of the law, rights and responsibilities are available and given to all at work. As Regulations made under the Act have the same scope, there now exists the potential to achieve clear and uniform standards.

An "enabling" Act, much of the text is devoted to the legal machinery for creating administrative bodies, combining others and detailing new powers of inspection and enforcement. The Health and Safety Commission (HSC) – see below – carries responsibility for policy-making and enforcement, answerable to the Minister of State for Employment. Its executive arm is the Health and Safety Executive (HSE), whose functions range from enforcement to research and European liaison on standards.

The gradual replacement of previous health and safety requirements by revised and updated measures applicable to the whole of the workforce of the country is provided for. This is done by the repeal of statutes and their replacement with Regulations and Approved Codes of Practice prepared in consultation with industry and workers. One of the key features of the Report of the Robens Committee, echoed by European Directives, is the principle of consultation at all levels in order to achieve consensus and combat apathy. This consultative process starts within the HSC, and continues to the workplace, where employers are required to consider the views of workers in the setting of health and safety standards.

The Act consists of four parts:

Part 1: Contains provisions on

- health and safety of people at work
- protection of others against health and safety risks from work activities
- control of danger from articles and substances used at work
- controlling certain atmospheric emissions

Part 2: Establishes the Employment Medical Advisory Service

Part 3: Amends previous laws relating to safety aspects of building regulations

Part 4: Contains a number of general and miscellaneous provisions.

GENERAL DUTIES OF EMPLOYERS

These are contained in **Sections 2, 3, 4 and 9**. By **Section 40**, the burden of proof is transferred from the prosecution to the defence in prosecutions where it is alleged that the accused person or employer failed to do what was practicable or reasonably practicable as required, in the particular circumstances. **Sections 36 and 37** provide for the personal prosecution of members of management in certain circumstances, notably in Section 37 where they can be charged as well as, or instead of, the employer if the offence in question was due to their consent, connivance or neglect.

Employers must, as far as is reasonably practicable, safeguard the health, safety and welfare of employees **(Section 2)**. In particular, this extends to the provision and maintaining of:

a) safe plant and safe systems of work

b) safe handling, storage, maintenance and transport of (work) articles and substances

c) necessary information, instruction, training and supervision

d) a safe place of work, with safe access and egress

e) a safe working environment with adequate welfare facilities

There is an absolute duty on employers with five or more employees to prepare and revise as necessary a written statement of safety policy, which details the general policy and the particular organisation and

The Health and Safety at Work Etc. Act 1974

arrangements for carrying it out. The policy must be brought to the notice of all the employees (**Section 2(3)**). This topic is discussed in Module C Section 4.

Employers must consult with employees on health and safety matters, and in particular with (trade union appointed) safety representatives. They must also set up a safety committee on request of the safety representatives with the main function of keeping under review the measures taken to ensure the health and safety at work of employees. (**Section 2 (4-7)**). Regulations made in 1977 expand employers' duties to consult with trade union appointees. These are the Safety Representatives and Safety Committees Regulations, which describe the functions of safety representatives but do not impose any duties upon them. The functions are: investigations of potential hazards and dangerous occurrences; examining the causes of accidents; investigation of complaints by employees; making representations to the employer on those matters; carrying out inspections of several kinds; representing employees in consultations with enforcing authorities; receiving information from inspectors and attending meetings of safety committees. A safety committee must be set up on the request of two or more safety representatives. Safety representatives have the right to training to carry out their functions; their entitlements are covered in an Approved Code of Practice and Guidance Notes. This topic is discussed in Module D Section 27.

The self-employed, other employees and the public must not be exposed to danger or risks to health and safety from work activities (**Sections 3, 4**).

Harmful emissions into the atmosphere must be prevented, from prescribed operations (**Section 5**).

GENERAL DUTIES OF THE SELF-EMPLOYED

Similar duties to the above rest upon the self-employed, with the exception of the safety policy requirement (unless they in turn employ others) (**Section 3 (2)**).

GENERAL DUTIES OF EMPLOYEES

By **Section 7**, employees must take reasonable care of their own health and safety and that of others who may be affected by their acts or omissions. They must also co-operate with their employer so far as is necessary to enable the employer to comply with his duties under the Act. By **Section 8**, it is an offence for anyone to intentionally or recklessly interfere with or misuse

anything provided in the interests of health, safety or welfare. Members of management, who are also employees, are vulnerable to prosecution under Section 7 if they fail to carry out their health and safety responsibilities (as defined in the safety policy statement), in addition to their liability under Sections 36 and 37 as noted above.

GENERAL DUTIES OF MANUFACTURERS AND SUPPLIERS

Section 6 of the Act focuses attention on the role of the producer of products used at work, and includes designers, importers and hirers-out of plant and equipment in the list of those who have duties under it. The Section refers to articles and substances for use at work, and requires those mentioned to ensure so far as is reasonably practicable that they are safe when being "used" in the widest sense. They must be tested for safety in use, or tests are to be arranged and done by a competent authority. Information about the use for which an article or substance was designed, including any necessary conditions of use to ensure health and safety, must be supplied with the article or substance.

Section 6 has been modified so as to apply to fairground equipment, to provide a more complete description of the kinds of use an article may be put to, and to ensure that necessary information is actually provided and not merely "made available" as was originally required.

Hire purchase companies are not regarded as suppliers for the purposes of Section 6. An interesting provision requires anyone installing an article for use at work to ensure, so far as is reasonably practicable, that nothing about the way in which it is installed or erected makes it unsafe or a risk to health when properly used.

CHARGES

Section 9 forbids the employer to charge his employees for any measures which he is required by statute to provide in the interests of health and safety.

THE HEALTH AND SAFETY EXECUTIVE AND COMMISSION

These bodies were established by the Act, in **Section 10**. Their chief functions are:

The **HSC** takes responsibility for developing policies in health and safety away from Government Departments.

The Health and Safety at Work Etc. Act 1974

Its nine members are appointed by the Government from industry, trade unions, local authorities and other interest groups (currently consumers). It can also call on Government Departments to act on its behalf.

The **HSE** is the enforcement and advisory body to the HSC. For lower risk activities and premises such as offices, the powers of enforcement are delegated to Local Authorities' Environmental Health Departments, whose Officers have the same powers as HSE Inspectors when carrying out these functions.

POWERS OF INSPECTORS

These are contained in **Sections 20-22, 24, 25, 33 and 42** of the Act. An appointed inspector can:

- Gain access without a warrant to a work place at any time
- Employ the police to assist in the execution of his duty
- Take equipment or materials onto premises to assist his investigations
- Carry out examinations and investigations as he sees fit
- Direct that locations remain undisturbed for as long as he sees fit
- Take measurements, photographs and samples
- Order the removal and testing of equipment
- Take articles or equipment away with him for examination or testing
- Take statements, records and documents
- Require the provision of facilities he needs to assist him with his enquiries
- Do anything else necessary to enable him to carry out his duties.

ENFORCEMENT

If an inspector discovers what he believes to be a contravention of any Act or Regulation he can:

- Issue a **prohibition notice**, with deferred or more usually immediate effect, which prohibits the work described in it if the inspector is of the opinion that the circumstances present a risk of serious personal injury, which is effective until the steps he may specify have been taken to remedy the situation. Appeal can be made to an industrial tribunal, but the notice remains in effect until the appeal is heard
- Issue an **improvement notice**, which specifies a time period for the rectification of the contravention of a statutory requirement. Appeal can be made to an industrial tribunal within 21 days; this has the effect of postponing the notice until its terms have been confirmed or altered by the tribunal
- Prosecute any person who contravenes a requirement or fails to comply with a notice as above
- Seize, render harmless or destroy any article or substance which he considers to be the cause of imminent danger or serious personal injury.

PENALTIES

Successful prosecutions under the Act or any health and safety statutes attract a maximum penalty of a fine of up to £5,000 and/or up to 6 months imprisonment. For breaches of the general duties Sections of the Act (2-6), penalties of up to £20,000 were introduced in early 1992, for those dealt with in the Magistrates' Court. Offences heard on indictment in the Crown Court attract unlimited financial penalties and up to 2 years imprisonment. The present single largest fine recorded at the Crown Court is £500,000 for a single incident (BP Grangemouth). Penalties for breaching prohibition notices are normally the largest, as are those resulting from fatalities. The financial consequences of receiving prohibition and improvement notices are often severe, because of work delays and disruption.

NEW MANAGEMENT REGULATIONS

The detailed provisions of the Management of Health and Safety at Work Regulations 1992 are discussed in the next Section. It has been claimed that their effect is solely to provide more detailed descriptions of the general duties owed under the Act, and as such will make little difference to employers already complying with the Act. However, it should be noted that there are several aspects of the Regulations which impose new duties on employers and others, particularly in respect of measures to deal with serious and imminent danger, information required to be given to others, and those requiring written assessments to be made.

The Health and Safety at Work Etc. Act 1974

SELF-ASSESSMENT QUESTIONS

1. What are the powers of an HSE Inspector? What enforcement action can be taken by an Inspector to prevent dangerous acts taking place?

2. Summarise the employer's duties under the Health and Safety at Work etc. Act 1974.

REVISION

The Health and Safety at Work etc. Act 1974 is an enabling Act in 4 Parts.

Part I sets out:

- general duties of employers, employees, self-employed and those involved in the supply process of anything used at work; established the HSE and HSC and their enforcement functions; provides for future Regulations and Approved Codes of Practice.

Part II establishes:

- the Employment Medical Advisory Service (EMAS)

Part III amends the Building Regulations

Part IV contains miscellaneous and general provisions.

The Management of Health and Safety at Work Regulations 1992

INTRODUCTION

The Regulations implement most of a European Directive No 89/391/EEC of 29th May 1990, the "Framework Directive", and also Council Directive 91/383/EEC dealing with the health and safety of those who are employed on a fixed term or other temporary basis. The Regulations are in addition to the requirements of the Health and Safety at Work etc. Act 1974, and they extend the employer's general safety obligations by requiring additional specific actions on the employer's part to enhance control measures.

NB. For the first time in Regulations (as opposed to the Health and Safety at Work etc. Act 1974 which also contains it), Regulation 15 provides that a breach of a duty imposed by these Regulations does not confer a right of action in any civil proceedings.

These Regulations do not cover all the matters mentioned in the Directives, partly because these are dealt with elsewhere, and partly because existing provisions are considered to cover the points adequately. Examples include first-aid, use of personal protective equipment, and fire-fighting.

The Regulations do not apply to the master or crew of a sea-going ship, or to their employer in respect of the normal ship-board activities of the crew.

There are some overlaps between this general legislation and specific Regulations such as the Manual Handling Operations Regulations 1992. Where these general duties are similar to specific ones elsewhere, as in the case of general risk assessments and those required by other specific Regulations, the legal requirement is to comply with both the general and the specific duty. However, specific assessments required elsewhere will also satisfy the general requirement under these (MHSW) Regulations for risk assessment **as far as those operations alone are concerned**. The employer will still have to apply the general duties in all work areas.

The **purpose of the Regulations** is to implement the Directives referred to above.

REQUIREMENTS OF THE REGULATIONS

Regulation 3 requires every employer and self-employed person to make a suitable and sufficient assessment of the health and safety risks to employees and others not in his employment to which his undertakings give rise, in order to put in place appropriate control measures. It also requires review of the assessments as appropriate, and for the significant findings to be recorded (electronically if desired) if five or more are employed. Details are also to be recorded of any group of employees identified by an assessment as being especially at risk.

Regulation 4 requires employers to make appropriate arrangements given the nature and size of his operations, for effective planning, organising, controlling, monitoring and review of the preventive and protective measures he puts in place. If there are more than five employees the arrangements are to be recorded. **Regulation 5** requires the provision by the employer of appropriate health surveillance, identified as being necessary by the assessments.

Adequate numbers of 'competent persons' must be appointed by employers under **Regulation 6**. They are to assist the employer to comply with obligations under all the health and safety legislation, unless (in the case only of sole traders or partnerships) the employer already has sufficient competence to comply without assistance. This Regulation also requires the employer to make arrangements between competent persons to ensure adequate co-operation between them, to provide the facilities they need to carry out their functions, and specified health and safety information (on temporary workers, assessment results, risks notified by other employers under Regulation 9(1)(c), and procedures for danger and danger areas required by Regulation 7).

Regulation 7 requires employers to establish and give effect to procedures to be followed in the event of serious and imminent danger to persons working in their undertakings, to nominate competent persons to implement any evacuation procedures and restrict access to danger areas. Persons will be classed as competent when trained and able to implement these procedures. For example, one effect of this Regulation is to require fire evacuation procedures with nominated fire marshals in every premises.

Regulation 8 requires employers to give employees comprehensible and relevant information on health and safety risks identified by the assessment, the protective and preventive measures, any procedures under Regulation 7, the identities of competent persons appointed under Regulation 7, and risks notified to the employer by others in shared facilities (Regulation 9).

Regulation 9 deals with the case of two or more employers or self-employed persons sharing a workplace. Each employer or self-employed person in this

The Management of Health and Safety at Work Regulations 1992

position is to co-operate with any others as far as necessary to enable statutory duties to be complied with, reasonably co-ordinate his own measures with those of the others, and take all reasonable steps to inform the other employers of risks arising out of his undertaking. This Regulation applies whether the workplace is shared on a permanent or temporary basis.

Regulation 10 requires employers and the self-employed to provide employers of any employees, and every employee from outside undertakings, and every self-employed person, who is working in their undertakings with comprehensible information concerning any risks from the undertaking. Outside employers, their employees and the self-employed also have to be told how to identify any person nominated by the employer to implement evacuation procedures under Regulation 7.

Regulation 11 requires employers to take capabilities as regards health and safety into account when giving tasks. Adequate health and safety training must be provided to all employees on recruitment, and on their exposure to new or increased risks because of: job or responsibility change, introduction of new work equipment or a change in use of existing work equipment, introduction of new technology, or introduction of a new system of work or a change to an existing one. The training must be repeated where appropriate, take account of new or changed risks to the employees concerned, and take place during working hours.

Regulation 12 introduces new employee duties. They must use all machinery, equipment, dangerous sub-stances, means of production, transport equipment and safety devices in accordance with any relevant training and instructions, and inform their employer or specified fellow employees of dangerous situations and short-comings in the employer's health and safety arrangements.

Regulation 13 requires employers and the self-employed to provide any person they employ on a fixed-term contract or through an employment business (an agency) with information on any special skills required for safe working and any health surveillance required, before work starts. The employment business (agency) must be given information about special skills required for safe working, and specific features of jobs to filled by agency workers where these are likely to affect their health and safety. The agency employer is required to provide that information to the employees concerned.

Exemptions can be granted by certificate from the Secretary of State for Defence for visiting or home forces (**Regulation 14**). **Regulation 16** extends the Regulations to premises and activities outside Great Britain to which the Health and Safety at Work etc. Act 1974 applies, and also extends the meaning of "at work" so that persons shall be treated as being at work throughout the time they are present at the premises outside Great Britain.

Regulation 17 modifies the Safety Representatives and Safety Committees Regulations 1977 (SRSC) as specified in the Schedule to the Regulations. (See Section 27).

SELF-ASSESSMENT QUESTIONS

1. Review your organisation's health and safety policy and make notes on improvements needed to comply with these Regulations.

2. Summarise the main ways in which these Regulations expand the duties of the employer under the Health and Safety at Work etc. Act 1974.

REVISION

Ten requirements of the Management of Health and Safety at Work Regulations:

- Risk assessments
- Formal management control systems
- Health surveillance
- Competent person appointments

- Procedures for serious and imminent danger
- Information for workers
- Inter-employer co-operation
- Job-specific training
- Capability assessment
- Further employee duties

The Workplace (Health, Safety and Welfare) Regulations 1992

INTRODUCTION

The Regulations implement most of the requirements of the Workplace Directive (89/654/EEC) concerning minimum standards for workplace health and safety. Most of the provisions have long been part of the British law, which has been, however, not comprehensive in application. As a result, much of the pre-1974 law in this field is repealed or revoked by these Regulations.

There is a distinction to be made between new and old workplaces, such that workplaces existing prior to 1st January 1993 are not subject to these Regulations until 1st January 1996. For these, provisions of earlier legislation will remain in full until 1996. For new workplaces taken into use after 1st January 1993 (even if built before then), and for modifications, conversions and extensions after that date, the Regulations take immediate effect from 1st January 1993. Some workplaces not previously subject to specific requirements are now in scope; these include schools and hospitals.

The major **repeals and revocations** are Sections 1-7, 18, 28, 29, 57-60 and 69 of the Factories Act 1961, and Sections 4-16 of the Offices, Shops and Railway Premises Act 1963. Twenty-five sets of Regulations and Orders, and parts of a further seventeen disappear.

The **objective of the Regulations** is to place obligations on employers and others in control of workplaces to reduce the risks associated with work in or near buildings. Opportunity has also been taken to rationalise the different, wide-ranging and detailed requirements in pre-1993 UK health and safety law, and to combine them all in one comprehensive set of Regulations.

SOME SIGNIFICANT DEFINITIONS

- **Traffic route** means a route for pedestrians, vehicles, or both – and includes stairs, fixed ladders, doors, ramps and loading bays.

- **Workplace** means any non-domestic premises or part of premises made available to any person as a place of work. It includes any place within the premises to which a person has access while at work, and means of access to or egress from the premises other than a public road. A modification, conversion or extension is part of a workplace only when completed.

- **Premises** means any place, including an outdoor place. Domestic premises are not within the scope of the Regulations, which as a result do not cover homeworkers.

REQUIREMENTS OF THE REGULATIONS

The application of these Regulations is very general, in contrast to the Factories Act and other legislation which they largely replace. They do not apply to a workplace inside a means of transport, workplaces where the only activities are construction activities as defined, mineral resource extraction or ancillary activities. Fishing boats are also excluded. By **Regulation 3(2)** the provisions of Regulations 20-25 as they affect temporary work sites are to be qualified by "reasonably practicable". **Regulation 3(3)** removes application of Regulations 5-12 and 14-25 from workplaces in or on most means of transport. Regulation 13 only applies when the means of transport is stationary inside a workplace or, for licensed vehicles, off the road.

By **Regulation 3(4)**, Regulations 5-19 and 23-25 do not apply to places of work in woods, fields and other agricultural land which are not inside buildings, and in the same way Regulations 20-22 are subject in those circumstances to the test of reasonable practicability.

Where construction work is in progress within a workplace it can be treated as a construction site and not subject to these Regulations only if it is fenced off, otherwise these Regulations and the Construction Regulations apply. For temporary work sites, Regulations 20-25 apply so far as is reasonably practicable only. These include work sites used infrequently or only for short periods, and fairs and similar structures only present for short periods.

Generally, **Regulation 4** obliges employers and others to ensure that every workplace under their control and where any of their employees work complies with any applicable requirement. The same duty is placed on any person having control of a workplace to any extent in connection with trade, business or other undertaking (whether or not for profit), in relation to matters within his control. The self-employed are not under a requirement by Regulation 4 in respect of their own work or that of a partner. Every person deemed to be a factory occupier by Section 175(5) of the Factories Act 1961 shall ensure that those premises comply with these Regulations.

Regulation 5 covers maintenance of workplaces, equipment, devices and systems, which are to be cleaned, maintained in an efficient state, working order and good repair. If a fault in any equipment or device is liable to result in a breach of any of these Regulations, it must be subject to a suitable system of maintenance. For example, this would apply to emergency lighting and

The Workplace (Health, Safety and Welfare) Regulations 1992

ventilation and would extend to arrangements to ensure its continuous and effective operation. These and other examples would be expected to be identified by the risk assessments made under the Management of Health and Safety at Work Regulations 1992 (MHSWR).

Ventilation is addressed specifically by **Regulation 6**, which requires every enclosed workplace to be ventilated by a sufficient quantity of fresh or purified air, and plant to achieve this must be properly maintained and include an effective device to give visible or audible warning of any failure where this is necessary for health or safety reasons. The only exception to this is any confined space, where use of breathing apparatus may be necessary.

A reasonable temperature during working hours in all workplaces inside buildings replaces specified temperatures in previous legislation (**Regulation 7**), together with a ban on heating or cooling methods which allow injurious or offensive fumes to enter a workplace. Suitable thermometers in sufficient numbers are to be provided and maintained for workers to establish the temperature in any workplace inside a building. The Approved Code of Practice gives a sedentary minimum temperature of 16 degrees Celsius, and 13 degrees where there is severe physical effort, unless lower maximum room temperatures are required by law. (Note, however that the general Food Hygiene Regulations do not specify maximum room temperatures).

Suitable and sufficient lighting is required by **Regulation 8** for every workplace, by natural light as far as is reasonably practicable. Additionally, where there is potential danger due to failure of any artificial light source, suitable and sufficient emergency lighting is required.

Cleaning and decoration requirements are given in **Regulation 9**. Floor, wall and ceiling surfaces inside buildings must be capable of being kept sufficiently clean. All workplaces and their furniture, furnishings and fittings must be kept clean, which includes keeping them free from any effluvia from drains or toilets, and keeping them free of waste materials apart from what is kept in suitable receptacles so far as is reasonably practicable.

Every room where persons work must have sufficient floor area, height and unoccupied space for purposes of health, safety or welfare (**Regulation 10**). Personal space is defined in the Approved Code of Practice as 11 cubic metres. There is a notional maximum ceiling height of 3 metres to be used in this calculation. For existing workplaces (factories, subject to the provisions of the Factories Act 1961, compliance with Part 1 of Schedule

1 to the Regulations (see below) will be deemed to satisfy the requirement.

By **Regulation 11**, workstations must be arranged to be suitable for any person who is likely to work there, and for any work likely to be done there. For those workstations outside buildings, there must be protection from adverse weather so far as is reasonably practicable, and so arranged that persons at the workstation are not likely to slip or fall. Also, the arrangement must enable anyone to leave it swiftly or receive assistance in an emergency. For each person at work in the workplace where a substantial part of the work must be done sitting, a suitable seat must be provided. This means suitable for the person as well as the task, with a suitable footrest where necessary.

Floors and traffic route surfaces must be constructed so that they are suitable for the purpose for which they are used (**Regulation 12**). It is specified that this means having no hole or slope, or being uneven or slippery, so as to expose persons to risks to health or safety. Floors must have effective means of drainage where necessary. For holes, no account is to be taken of one where adequate measures have been taken to prevent falls; for slopes, account is to be taken of any handrails provided, in deciding if there is exposure to risk to health or safety. Except where a traffic route would be obstructed, suitable and sufficient handrails and guards must be provided.

So far as is reasonably practicable, suitable and effective safeguards are required against any person falling a distance likely to cause personal injury, or being struck by a falling object likely to cause personal injury. **Regulation 13** adds that any area where there is a risk to health and safety from either of these must be clearly indicated as appropriate. So far as is reasonably practicable, every tank, pit or structure shall be securely covered or fenced where there is a risk of falling into a dangerous substance in it. Also, every traffic route over these must be securely fenced. 'Dangerous substances' are defined as those which are likely to burn or scald, and those which are poisonous, corrosive or as fumes, gases or vapours are likely to overcome a person. Also included are any granular or free-flowing solid, or viscous substance, of a nature or quantity likely to cause danger to anyone.

Windows, doors, gates, walls and partitions glazed wholly or partially so that they are transparent or translucent must, where necessary for health or safety, be of safe material and appropriately marked (**Regulation 14**). By **Regulation 15**, windows, skylights and ventilators which can be opened are required to be designed so that they cannot be opened, closed or adjusted in a way which causes danger to anyone. They are also not to be capable of remaining open in a dangerous position. Windows and skylights are to be designed or constructed so that they can be cleaned safely, taking account of equipment or devices fitted to the window, skylight or building (**Regulation 16**).

The safe circulation of pedestrians and vehicles in the workplace must be organised, to comply with **Regulation 17**. Traffic routes are to be suitable for the use made of them, sufficient, in the right places, and big enough. This means that traffic routes must be so arranged that vehicles using them do not endanger those at work nearby, there is sufficient separation of vehicles and pedestrians at doors and gates, and where both use the same route there is sufficient separation between them. All traffic routes are to be signed where necessary for health and safety. The requirements on traffic routes (Regulation 17(2) and (3)) only apply so far as is reasonably practicable to a pre-1993 workplace.

Doors and gates (**Regulation 18**) must be suitably constructed, to include fitting with any necessary safety device. This means particularly that sliding and powered doors cannot cause injury by falling on or trapping people. Powered doors and gates require suitable features to prevent injury by trapping anyone, and should be operable manually unless they open automatically if the power fails. Any door or gate which can be pushed open from either side must provide a clear view of both sides of the space when closed.

Regulation 19 requires travelators and escalators to function safely, be equipped with any necessary safety devices and have one or more identifiable and accessible emergency stop control.

Sanitary conveniences must be suitable, sufficient, and in readily accessible places for all persons at work. The former means, according to **Regulation 20**, that the rooms containing them must be adequately lit and ventilated, rooms and conveniences are to be kept clean and orderly, and separate rooms containing conveniences being provided for men and women. The last is **not** required where a convenience is in a room intended for use by one person at a time and which has a door that can be secured from inside. For existing factories (pre-1993), compliance with the former Section 7 of the Factories Act 1961 in terms of numbers of conveniences provided is deemed to satisfy these requirements of the Regulations.

Washing facilities which are suitable and sufficient are required for all persons at work in any workplace, and are to be readily accessible. This also includes showers if required for health or work reasons. **Regulation 21** defines "suitable" as those which:

- are provided in the immediate vicinity of every sanitary convenience (whether or not provided elsewhere as well)
- are provided in the vicinity of any changing rooms required by these Regulations (whether or not provided elsewhere as well)
- include a supply of clean hot and cold, or warm, water which is running water as far as practicable
- include soap or other suitable cleanser
- include towels or other suitable means of drying
- are contained in rooms sufficiently ventilated and lit
- are together with their rooms kept clean and orderly, and
- provide separate facilities for men and women, unless they are in a room which can only be used by one person at a time and where the door can be secured from the inside.

The final requirement, for separate facilities, does not apply where they are used only for washing the face, hands and forearms.

An adequate supply of wholesome drinking water is to be provided and maintained for all persons at work in the workplace, which is to be readily accessible in suitable places and conspicuously marked by a sign where this is necessary for health and safety. An example would be where people could otherwise drink from a contaminated water supply. A sufficient number of drinking vessels is also required by **Regulation 22**, unless a water jet is installed where persons can drink easily.

Accommodation for clothing which is suitable and sufficient is required for personal clothing not worn during work hours, and for special clothing worn at work but not taken home (**Regulation 23**). "Suitable" accommodation is that which: where changing facilities are required by Regulation 24, is secure for clothes not worn; where necessary to avoid risks or damage to the clothing it includes separate facilities for work clothes and for other clothes; allows or includes drying facilities so far as is reasonably practicable; and is in a suitable location. **Regulation 24** requires suitable and sufficient changing facilities in all cases where special work clothing has to be worn and a person cannot be expected to change elsewhere for reasons of health or propriety. "Suitable" facilities will include separate facilities or separate use of facilities by men and women where necessary for propriety.

Suitable and sufficient rest facilities are to be provided at readily accessible places (**Regulation 25**). For new (post 1992) workplaces, conversions or extensions, the facilities must include one or more rest rooms. For

The Workplace (Health, Safety and Welfare) Regulations 1992

pre-1993 workplaces, the facilities can be either rest rooms or rest areas. They must also include facilities to eat meals where food eaten in the workplace would otherwise become contaminated. Both rest rooms and rest areas must include suitable arrangements to protect non-smokers from discomfort caused by tobacco smoke. Suitable facilities must be provided for any working pregnant women or nursing mothers to rest. Suitable and sufficient facilities are to be provided for persons at work to eat meals where meals are regularly eaten in the workplace.

Exemption certificates may be issued by the Secretary of State for Defence for armed forces (**Regulation 26**). **Schedule 1** to the Regulations only applies to Regulations 10 and 20, giving leave for standards contained in the Factories Act 1961 (on overcrowding and numbers of sanitary conveniences) to continue to apply to factories which are not new workplaces, extensions or conversions, that is those which are pre-1993.

SELF-ASSESSMENT QUESTIONS

1. Consult the Approved Code of Practice to establish the minimum number of toilets for each sex required in your workplace.

2. Write notes on the application of the Regulations to a forestry worker in a wood.

REVISION

Fourteen major topics in the Workplace (Health, Safety and Welfare) Regulations:

- Maintenance of workplace and equipment servicing it
- Ventilation, temperature and lighting
- Cleanliness
- Workspace allocation
- Workstation design and arrangement
- Traffic routes and floors
- Fall protection
- Glazing
- Doors and gates
- Travelators and escalators
- Sanitary and washing facilities
- Drinking water supply
- Accommodation for clothing
- Facilities for changing, rest and meals.

The Provision and Use of Work Equipment Regulations 1992

INTRODUCTION

PUWER applies to both new and old equipment, although in different timescales. The general requirements contained in Regulations 1-10 apply from 1st January 1993 to all equipment. The more specific requirements of Regulations 11-24 apply from the 1st January 1993 for work equipment first provided after that date, but from 1st January 1997 for work equipment provided before it. This will have the effect that existing legislation (listed in Schedule 2 of the Regulations for repeal or revocation) will continue to operate alongside the new Regulations until 1st January 1997.

Equipment sold on as second-hand becomes classed as 'new equipment', and will have to meet the requirements of all the Regulations but not those 'essential safety requirements' specified in the Machinery (Safety) Regulations 1992. The same applies to hired and leased equipment.

In time, the Regulations will replace, in full or in part, seventeen codes of Regulations (including requirements of the Abrasive Wheels 1970, the Woodworking Machines Regulations 1974 and the Construction (General Provisions) Regulations 1961), seven Sections of the Factories Act 1961 (including old favourites, Sections 12-16), one Section of the Offices, Shops and Railway Premises Act 1963 (17) and two Sections of the Mines and Quarries Act 1954 (notably 81(1) and 82).

OBJECTIVES OF THE REGULATIONS

The Regulations implement the requirements of the EC Directive on the 'minimum health and safety requirements for the use of work equipment at the workplace' (generally known as the Use of Work Equipment Directive or UWED). They set objectives rather than establish prescriptive requirements, and simplify and clarify previous legislation which will be repealed eventually. The Regulations provide a coherent set of requirements ensuring the provision of safe work equipment and its safe use, irrespective of age or place of origin.

SOME SIGNIFICANT DEFINITIONS

- **Work equipment** – includes any machinery, appliance, apparatus or tool and any assembly of components which, in order to achieve a common end, are arranged and controlled so that they function as a whole. Examples include: dumper trucks, trench sheets, overhead projector, scalpel. Not included in the definition: livestock, substances, private cars, structural items.

- **Use** – in relation to work equipment, this means any activity involving work equipment. This includes starting, stopping, erecting, installing, dismantling, programming, setting, using, transporting, repairing, modifying, maintaining, servicing and cleaning.

- **Danger zone** – means any zone in or around machinery in which a person is exposed to a risk to health and safety from contact with a dangerous part of machinery or a rotating stock-bar.

REQUIREMENTS OF THE REGULATIONS

Regulations 1-4 contain citation, definitions and application. Almost all employment relationships and the places of work involved, where the Health and Safety at Work etc. Act 1974 applies, are within scope. This includes all industrial sectors, offshore operations and service occupations. All activities involving work equipment are dealt with in the Regulations, including stopping, starting, use, transport, repair, modification, maintenance, servicing and cleaning. Any machine, appliance, apparatus or tool used at work or made available for use in non-domestic premises is covered. All the requirements apply to employers, the self-employed in respect of personal work equipment, persons holding obligations under Section 4 of the Health and Safety at Work etc. Act 1974 (control of premises) in connection with the carrying on of a trade, business or other undertaking, and persons who are occupiers of factories as defined by Section 175 of the Factories Act 1961.

Regulation 5 requires employers and others to ensure that work equipment is suitable for the purpose for which it is used. Selection of work equipment must have regard to working conditions and any additional risks posed by the use of the work equipment. The equipment must be used only for operations for which it is suitable. **Regulation 6** requires the equipment to be properly maintained, and have an up-to-date maintenance log where one is kept. Specific risks are dealt with by **Regulation 7**, which restricts use of equipment to persons given the task of using it, and where maintenance or repairs have to be done, those doing it must be nominated.

Users and supervisors of equipment must be given adequate health and safety information, and where appropriate, specified written instructions relating to

The Provision and Use of Work Equipment Regulations 1992

the use of work equipment (**Regulation 8**). User and supervisor training, including work methods, risks and precautions are covered in **Regulation 9**.

All work equipment taken into use for the first time after 31st December 1992 will have to comply with any other relevant Community Directives (**Regulation 10**), for example, the Machinery Safety Directive (89/392/EEC), and those on simple pressure vessels, tractors, noise, electromedical equipment and industrial trucks.

Regulations 11-24 deal with specific hazards associated with the use of work equipment. In most cases the requirements are aimed at the **provision** of equipment which is safe and without risks to health and the need to ensure that work equipment is provided with appropriate safety devices or protected against failure. In other cases, the requirements affect the **use** of work equipment.

Specific requirements are designed to reduce the risk to employees from dangerous parts of machinery (**Regulation 11**). This includes measures to prevent access to dangerous parts of machinery and to stop movement of any dangerous part before someone (or part of them) enters the danger zone. These measures must consist of guards or protection devices as far as practicable, and detailed requirements relating to them are also established. In working out the existence, size and position of a danger zone, account is only taken of the risk of contact with dangerous parts of machines.

In selecting preventive measures, the Regulation sets out a hierarchy of them on four levels. these are:

- Fixed, enclosing guards to the extent practicable, but where not –
- Other guards or protection devices to the extent practicable, but where not –
- Protection appliances (jigs, push-sticks etc) to the extent practicable, but where not –
- Provision of information, instruction, training and supervision.

Guards and devices are to be:

- Suitable for the purpose
- Of good construction, sound material and adequate strength
- Adequately maintained, in good repair and efficient working order
- Not the source of additional risk to health and safety
- Not easily bypassed or disabled
- Situated at sufficient distance from the danger zone
- Not unduly restrictive of any necessary view of the machine

- Constructed or adapted to allow maintenance or part replacement without removing them.

Exposure of a person to specified hazards must be prevented by the employer as far as reasonably practicable, or adequately controlled where it is not (**Regulation 12**). The protection is not to be by provision of personal protective equipment or information, training and supervision so far as is reasonably practicable, and is to include measures to minimise the effects of the hazard as well as to reduce the risk. The specified hazards are:

- Ejected or falling objects
- Rupture or disintegration of parts of the work equipment
- Fire or overheating of the work equipment
- The unintended or premature discharge or ejection of any article or of any gas, dust, liquid, vapour or other substance produced, used or stored in the work equipment
- The unintended or premature explosion of the work equipment or any material produced,produced, used or stored in it.

This Regulation does not apply where other specified Regulations apply, on COSHH, noise, ionising radiations, asbestos, lead and head protection in construction activities.

Measures must be taken to ensure that people do not come into contact with work equipment, parts of work equipment or any article or substance produced, used or stored in it which are likely to burn, scald or sear (**Regulation 13**).

Specific requirements relate to the provision, location, use and identification of control systems and controls on work equipment (**Regulations 14-18**). They relate to controls for starting or making a significant change in operating conditions, stop controls, emergency stop controls, controls in general and control systems. Several of these requirements, whilst good practice, are new to the legislative field.

All work equipment is to have a means to isolate it from all its source of energy (**Regulation 19**). The means will have to be clearly identifiable and readily accessible, and reconnection of equipment to any energy source must not expose people to any risk to health or safety. Work equipment is to be stable by clamping or otherwise where necessary, so as to avoid risks to health or safety (**Regulation 20**). Places where work equipment is used have to be suitably and sufficiently lit, taking account of the kind of work being done (**Regulation 21**).

As far as is reasonably practicable, maintenance operations are to be done while the work equipment is stopped. Otherwise, other protective measures are to be taken, unless maintenance people can do the work without exposure to a risk to health or safety **(Regulation 22)**. Employers must ensure that all work equipment has clearly visible markings where appropriate **(Regulation 23)** and any warnings or warning devices appropriate for health and safety **(Regulation 24)**. Warnings will be inappropriate unless they are unambiguous, easily perceived and easily understood.

SELF-ASSESSMENT QUESTIONS

1. You intend to buy a new abrasive wheel machine for use at work. What checks would you make to ensure it complies with these Regulations?

2. What extra steps must be taken to ensure compliance when the machine is taken into use?

REVISION

Ten requirements of the Provision and Use of Work Equipment Regulations:

- Suitability
- Maintenance
- Use by persons given the task
- Information

- Training
- 'CE' conformity
- Risk reduction
- Control improvements
- Isolation arrangements
- Warnings.

The Manual Handling Operations Regulations 1992

INTRODUCTION

Significant obligations are placed on employers and the self-employed with regard to all manual handling tasks in the workplace. The Regulations are the result of the European Directive on minimum health and safety requirements for the manual handling of loads where there is a risk particularly of back injury to workers [90/269/EEC].

The Regulations replace all former provisions on the subject, and apply to all those at work regardless of their particular workplace. Previous legislation has proved to be inadequate, being only applicable to certain work industries and difficult to interpret and enforce. It employed vague terms related to single lifts of loads; some Regulations covering particular industries applied load limits, again for single loads. The Regulations concentrate attention on ergonomic solutions, and take account of a range of relevant factors including repetitive lifting. Upper limb disorders are not addressed by the Regulations.

A clear step-by-step approach to removing hazards and minimising risks is introduced. The most fundamental requirement is to avoid hazardous manual handling operations so far as is reasonably practicable. Automation, redesigning the task to avoid manual movement of the load and mechanisation have to be considered at this point, if there is a risk of injury and a manual handling operation (as defined) is involved. If risk remains, a suitable and sufficient assessment of the operation has to be made, and that risk has to be reduced so far as is reasonably practicable, preferably involving mechanical assistance.

The **objectives of the Regulations** are to implement the Directive and apply an ergonomic approach to the prevention of injury while carrying out manual handling tasks.

SOME IMPORTANT DEFINITIONS

"Injury" – includes those resulting from the weight, shape, size, external state, rigidity (or lack of rigidity) of the load or from the movement or orientation of its contents. It does not include those resulting from the spillage or leakage of the contents (i.e those areas already controlled by other Regulations such as the COSHH Regulations).

"Load" – includes any person and any animal. Generally, this would be a discrete, movable object and would include a patient receiving medical attention and an animal undergoing veterinary treatment. This definition also covers material supported on a shovel or fork. However, an implement, tool or machine such as a chainsaw is not included.

"Manual Handling Operation" – means any transporting or supporting of a load (including lifting, putting down, pushing, pulling, carrying or moving) by hand or bodily force (such as using the shoulder). The force applied must be human and not mechanical.

REQUIREMENTS OF THE REGULATIONS

The Regulations impose the same duties on the self-employed as employers (**Regulation 2(2)**) but do not apply to normal activities on sea-going ships under the direction of the master (**Regulation 3**).

Regulation 4 places duties on employers to make evaluations and then assessments of certain manual handling operations. They must, so far as is reasonably practicable, avoid the need for employees to carry out those operations which involve a risk of injury (**Regulation 4(1)(a)**). Where this cannot be done, **Regulation 4(1)(b)** requires them to:

- make and keep up to date a suitable and sufficient assessment of all such manual handling tasks (**4(1)(b)(i)**), considering the factors and questions specified in the **Schedule** to the Regulations. Assessments are to be reviewed and amended if appropriate – if they become invalid or there is a significant change in the manual handling operation to which they relate
- take appropriate steps to reduce the risk of injury to employees arising from any such operation to the lowest level reasonably practicable (**4(1)(b)(ii)**), and
- take appropriate steps to provide employees who are carrying out manual handling operations with general indications and, where reasonably practicable to do so, precise information on the weight of each load and the heaviest side of any load whose centre of gravity is not centrally positioned (**4(1)(b)(iii)**).

Regulation 5 requires employees to make full and proper use of any system of work put in place by the employer to reduce the risk of injury during manual handling. This duty is in addition to the general duties of employees under Regulations 7 and 8 of the Health and Safety at Work etc. Act 1974.

Regulation 6 provides for the issue of exemption certificates by the Secretary of State for Defence for armed forces in the interests of national security. **Regulation 7** extends the requirements to offshore in territorial waters. **Regulation 8** revokes all the previous specific load-limiting Regulations, and Sections of the Factories Act 1961 and the Offices, Shops and Railway Premises Act 1963 referring to prohibitions on 'loads so heavy as to be likely to cause injury'.

NB. These requirements must be taken in conjunction with other duties imposed, particularly by the Management of Health and Safety at Work Regulations, where the reader's attention is drawn to the need for capability assessment and training. Module A Section 3 gives more information on the practical implications of manual handling, and information on risk assessment is available in Module C Section 3.

SELF-ASSESSMENT QUESTIONS

1. Identify activities in your workplace where these Regulations apply? List them.

2. Write notes on how these requirements affect a company supplying and installing double glazing.

REVISION

Five requirements of the Manual Handling Operations Regulations:

- Avoidance of manual handling
- Mechanise or automate process
- Assessment
- Risk reduction
- Information.

The Health and Safety (Display Screen Equipment) Regulations 1992

INTRODUCTION

The Regulations implement a European Directive (90/270/EEC) on minimum health and safety requirements for work display screen equipment (DSE). The Regulations do not replace general obligations under the Health and Safety at Work etc. Act 1974, the Management of Health and Safety at Work Regulations 1992, the Workplace (Health, Safety and Welfare) Regulations 1992 and the Provision and Use of Work Equipment Regulations 1992.

There are some overlaps between general and specific legislation such as this. Where general duties are similar to specific ones, as in the case of general risk assessments and those required by these Regulations, the legal requirement is to comply with both the general and the specific duty. However, assessment of workstations required by Regulation 2 will also satisfy the general requirement under the Management (MHSW) Regulations for risk assessment as far as those workstations are concerned. The employer still has to apply the general duties in other work areas, and other parts of work areas where display screen work is being carried out.

The **objectives of the Regulations** are to improve working conditions at display screen equipment by providing ergonomic solutions, to enable certain regular users of the equipment to obtain eye and eyesight tests and information about hazards, risks and control measures associated with their workstations.

POSSIBLE HEALTH EFFECTS OF DSE WORK

Over the past twenty years, DSE has been claimed to be responsible for a wide variety of adverse health effects, including radiation damage to pregnant women. Medical research shows, however, that radiation levels from the equipment do not pose significant risks to health. The range of symptoms which is positively linked relates to the visual system and working posture, together with general increased levels of stress and fatigue in some cases.

Repetitive strain injury (RSI) arising from work activities is now known with other musculoskeletal problems as work related upper limb disorders (WRULDS). They range from temporary cramps to chronic soft tissue disorders such as carpal tunnel syndrome in the wrist. It is likely that a combination of factors produces these. They can only be prevented by an analysis of the workstation as required by the Regulations, together with an appreciation of the role of training, job design and work planning.

Eye and eyesight defects do not result from use of DSE, and it does not make existing defects worse. Temporary fatigue, sore eyes and headaches can be produced by poor positioning of the DSE, poor legibility of screen or source documents, poor lighting and screen flicker. Staying in the same position relative to the screen for long periods can have the same effect.

Fatigue and stress are more likely to result from poor job design, work organisation, lack of user control over the system, social isolation and high-speed working than from physical aspects of the workstation. These factors will also be identified in an assessment, and its associated consultation with the workforce.

Epilepsy is not known to have been induced by DSE. Even photosensitive epileptics (who react to flickering light patterns) can work safely with display screens.

Facial dermatitis has been reported by some DSE users, but this is quite rare. The symptoms may be due to workplace environmental factors including low humidity and static electricity near the equipment.

Pregnancy is not put at risk by DSE work, but those worried about the dangers should be encouraged to talk with someone aware of current advice. This is that DSE radiation emissions do not put unborn children at risk – or anyone else.

SOME IMPORTANT DEFINITIONS

Display screen equipment – any alphanumeric or graphic display screen, regardless of the display process involved. This includes CRT display screens, liquid crystal and other new technologies. Screens showing mainly TV or film pictures are not covered. Microfiche screens are included in the definition. **Regulation 1(4)** states the following are not covered by these Regulations: window typewriters, calculators, cash registers and other equipment with small data displays, portable systems not in prolonged use, display screen equipment mainly intended for public use (cashpoints etc), on a means of transport and in drivers' or control cabs for vehicles or machinery.

Use – means in connection with work.

User – an employee who habitually uses display screen equipment (DSE) as a significant part of his or her normal work.

Operator – a self-employed person who habitually uses display screen equipment as a significant part of his or her normal work.

The Health and Safety (Display Screen Equipment) Regulations 1992

Workstation – the immediate work environment around the display screen equipment, including all accessories, desk, chair, keyboard, printer and other peripheral items.

MORE ABOUT USERS AND OPERATORS

An individual will generally be classified as a user or operator if most or all of the following apply:

1. Depends on the use of DSE to do the job (no alternative means readily available to achieve the same results)

2. Has no discretion as to use or non-use of the DSE

3. Needs significant training and/or special skills in the use of DSE to do the job

4. Normally uses DSE for continuous spells of an hour or more at a time

5. Uses DSE as above more or less daily

6. Job requires fast transfer of information between user and screen

7. Performance requirements of the system demand high levels of attention and concentration.

Some examples of those likely to be classified as users/operators are: WP pool workers, personal secretaries, data input operators, journalists, tele-sales and enquiry operators, librarians. If there is doubt, carrying out a risk assessment should assist in making the decision. The guidance material accompanying the Regulations contains a helpful discussion and examples of this classification.

REQUIREMENTS OF THE REGULATIONS

Regulation 2 requires every employer to perform a suitable and sufficient analysis of those workstations which are used by users or operators, to assess the health and safety risks to which they are exposed as a consequence. An assessment is to be reviewed by the employer if there has been a significant change in the matters to which it relates, or if the employer suspects it is no longer valid. The employer is then required to reduce the risks so identified to the lowest extent which is reasonably practicable. "Significant change" includes a major change in the software used, the hardware,

furniture, increase in time spent using the DSE, increase in task requirement such as speed and accuracy, relocation of the workstation and modification to the lighting.

Regulation 3 requires that new workstations used by users or operators must meet the requirements of the Schedule to the Regulations. Existing (pre-1993) workstations must comply by not later than 31st December 1996. The **Schedule** to the Regulations lists minimum requirements for the workstation; the matters dealt with are:

- Display screen
- Keyboard
- Work desk/surface
- Work chair
- Space requirements
- Lighting
- Reflections and glare
- Noise
- Heat
- Radiation (but no action is necessary)
- Humidity
- Computer/user-operator interface.

Regulation 4 requires the planning of activities of users by the employer so that daily work is periodically interrupted by breaks or activity changes. General guidance is given that these could be informal breaks away from the screen for a short period each hour.

Regulation 5 gives users or those about to become users the opportunity to have an appropriate eye and eyesight test as soon as practicable after request and at regular intervals. This entitlement also applies when a user experiences visual difficulties attributable to DSE work. Special corrective appliances are to be provided by the employer for users where normal ones cannot be used. ('Special' appliances will be those prescribed to meet vision defects at the viewing distance – anti-glare screens are not special corrective appliances). The costs of tests and special corrective appliances are to be met by the employer of the user.

Regulation 6 requires the user's employer to provide adequate health and safety training in the use of any workstation he may be required to work on, and further training whenever the organisation of any such workstation is substantially modified. The employer is to provide operators and users with information about risk assessments and control measures concerning the health and safety aspects of their workstations (**Regulation 7**). Users are to be given information about breaks and activity changes, eye and eyesight tests, and training both initially and when the workstation is modified.

The Health and Safety (Display Screen Equipment) Regulations 1992

The Secretary of State for Defence can issue exemption certificates for armed forces if in the national interest to do so (**Regulation 8**). The Regulations are extended outside Great Britain in parallel with the application of the Health and Safety at Work etc. Act 1974 (**Regulation 9**).

SELF-ASSESSMENT QUESTIONS

1. List the equipment in your work area to which these Regulations apply, and identify any 'users'.

2. Explain the difference between a 'user' and an 'operator'.

REVISION

Five requirements of the Health and Safety (Display Screen Equipment) Regulations are:

- Risk assessment
- Schedule compliance
- Work breaks/activity changes
- Eye and eyesight tests, corrective appliances
- Training and information.

Personal Protective Equipment at Work Regulations 1992

INTRODUCTION

The Personal Protective Equipment at Work Regulations replace much of the prescriptive requirements of legislation made prior to the Health and Safety at Work etc Act 1974, and implement the requirements of a European Directive. They place duties on both employers and the self-employed. There have long been piecemeal requirements in various statutes, but these have now been repealed or revoked and the Regulations provide a framework for the provision of personal protective equipment (PPE) in circumstances where assessment has shown a continuing need for personal protection.

PPE used for protection while travelling on a road, such as cycle helmets, crash helmets and motorbike leathers worn by employees on the highway is not covered by the Regulations. However, the requirements do apply in circumstances where this type of equipment is worn elsewhere (other than the highway) at work. An example is the wearing of crash helmets by farm workers using all-terrain vehicles. Protective equipment used during the playing of competitive sports is not covered either. Self-defence equipment (personal sirens/alarms) or portable devices for detecting and signalling risks and nuisances (such as personal gas detection equipment or radiation dosimeters) are not within the scope of the Regulations. Offensive weapons are not items of personal protective equipment!

The requirements of the Regulations do not apply to most respiratory protective equipment, ear protectors and some other types of PPE because they are already covered by existing Regulations such as COSHH, Ionising Radiations Regulations, Control of Lead and Asbestos at Work Regulations and the Construction (Head Protection) Regulations (See later Sections in this Module). However, this legislation is amended in varying degrees by Schedule 1 of the Regulations, which introduces the concept of risk assessment where previously missing. The Regulations also amend earlier provisions by requiring equipment to comply with European conformity standards and inserting a requirement to return it to suitable storage after use.

The **objectives of the Regulations** are to implement the Directive and formalise the provision of PPE following the assessment of risks required in companion Regulations.

PPE – THE DEFINITION

- **Personal protective equipment** – means all equipment designed to be worn or held by a person at work to protect against one or more risks, and any addition or accessory designed to meet this objective.

Both protective clothing and equipment are within the scope of the definition, and therefore such items as diverse as safety footwear, waterproof clothing, safety helmets, gloves, high visibility clothing, eye protection, respirators, underwater breathing apparatus and safety harnesses are covered by the Regulations. Ordinary working clothes or clothing provided which is not specifically designed to protect the health and safety of the wearer is not within the definition – such as clothes provided with the primary aim of presenting a corporate image and protective clothing provided in the food industry for hygiene purposes.

REQUIREMENTS OF THE REGULATIONS

The Regulations do not apply to the crews of ships while operational (**Regulation 3**). Shore-based contractors working on ships will be within scope of the Regulations whilst inside territorial waters. The requirements apply to PPE used on ships operating on inland waterways and aircraft while on the ground and operating in UK airspace.

Regulation 4 requires every employer to provide suitable PPE to each of his employees who may be exposed to any risk while at work, except where any such risk has been adequately controlled by other means which are equally or more effective. A similar provision applies to the self-employed.

PPE is to be used as a last resort. Steps should first be taken to prevent or control the risk at source by making machinery or processes safer and by using engineering controls and systems of work. Risk assessments made under the Management of Health and Safety at Work Regulations 1992, will help determine the most appropriate controls.

'Suitable' means:

a) appropriate for the risks involved and the conditions

b) takes account of ergonomic requirements and the state of health of the person wearing it

c) it is capable of fitting the wearer correctly after adjustments

d) it is effective to prevent or adequately control the risk without leading to any increased risk (particularly where several types of PPE are to be worn) so far as is practicable; and

Personal Protective Equipment at Work Regulations 1992

e) complies with national and European conformity standards in existence at the particular time. This requirement does not apply to PPE obtained before the harmonising Regulations come into force, which can continue to be used without European Conformity marks, but the employer must ensure that the equipment he provides remains 'suitable' for the purpose to which it is put.

Regulation 5 requires the employer and others to ensure compatibility of PPE in circumstances where more than one item of equipment is required to control the various risks.

Before choosing any PPE the employer or the self-employed person must make an assessment to determine whether the proposed PPE is suitable (**Regulation 6**). This assessment of risk is necessary because no PPE provides 100% protection and as with other assessments, judgment of types of hazard and degree of risk will have to be made to ensure correct equipment selection. The assessment must include:

a) an assessment of any risks which have not been avoided by any other means

b) the definition of the characteristics which PPE must have in order to be effective against the risks, including any risks which the PPE itself may create, and

c) a comparison of the characteristics of the PPE available with those referred to in (b).

For PPE used in high-risk situations or for complicated pieces of PPE (such as some diving equipment) the assessment should be in writing and be kept available. Review of it will be required if it is suspected to be no longer valid, or the work has changed significantly.

Regulation 7 requires all PPE to be maintained, replaced or cleaned as appropriate. Appropriate accommodation must be provided by the employer or self-employed when the PPE is not being used (**Regulation 8**). "Accommodation" includes pegs for helmets and clothing, carrying cases for safety spectacles and containers for PPE carried in vehicles. In all cases, storage must protect from contamination, loss or damage (particularly from harmful substances, damp or sunlight).

Regulation 9 requires an employer who provides PPE to give adequate and appropriate information, instruction and training to enable those required to use it to know the risks the PPE will avoid or limit, and the purpose, manner of use and action required by the employee to ensure the PPE remains in a fit state, working order, good repair and hygienic condition. Information and instruction provided must be comprehensible.

By **Regulation 10**, every employer providing PPE must take all reasonable steps to ensure it is properly used. Further, Regulation 10 requires every employee provided with PPE to use it in accordance with the training and instruction given by the employer. The self-employed must make full and proper use of any PPE. All reasonable steps must be taken to ensure the equipment is returned to the accommodation provided for it after use.

Employees provided with PPE are required under **Regulation 11** to report any loss or defect of that equipment to their employer forthwith. **Regulation 12** permits the granting of exemption certificates by the Secretary of State for Defence in respect of the armed forces. **Regulation 13** extends the applicability of the Regulations to all work activities in territorial waters and elsewhere subject to the Health and Safety at Work etc. Act 1974.

NB. By virtue of Section 9 of the Health and Safety at Work etc. Act 1974, no charge can be made to the worker for the provision of PPE which is used only at work.

SELF-ASSESSMENT QUESTIONS

1. Identify and list the items of personal protective equipment provided in your workplace.

2. Write notes on the duties of employers under the Regulations.

REVISION

Eight requirements of the Personal Protective Equipment at Work Regulations:

- Provision as 'the last resort'
- Suitability assessment
- Compatibility between items
- Maintenance and replacement
- Accommodation
- Information, instruction and training
- Proper use
- Report of loss/defect.

INTRODUCTION

Since the early part of the nineteenth century, when more and more people began to earn their living in factories and workshops instead of agriculture and handcraft work, Parliament has become correspondingly active in passing laws to promote health and safety at work. The first law was passed in 1802, and was promoted by pioneer reformers attempting to improve the conditions of child labour. This was "An Act for the preservation of the Health and Morals of Apprentices and others employed in Cotton and other Mills and other Factories." This Act was important, not only for the sanitary improvements it introduced, but also because it influenced further legislation and its direction. From then onwards until 1974, the law was aimed at types of **premises** (factories, shops, mines for example), and where necessary at types of work activity or **processes** going on there – grinding of metals, the use of special types of machinery, etc.

No further factory safety legislation was passed until 1819, but between then and 1856 there was a steady stream of statutes and subordinate Regulations covering areas such as safety, hours of work, the employment of women and children, and other important issues. By 1875, factory safety law was contained in numerous documents with no real pattern of development. Then, under a review by a Royal Commission, the Factory and Workshops Act was passed in 1878 which attempted to provide the first comprehensive piece of factory legislation. However, even this new system required reviews at intervals, to cope with changing technology and developing legal precedents, and further extensions of this Act were made in 1883, 1889, 1891, 1895 and 1897. In 1901 the Factory and Workshop Act became the principal statute controlling safety in factories, and remained so until its repeal by the Factories Act of 1937.

The 1937 Act provided for the first time a comprehensive code for safety, health and welfare requirements, applicable to all factories. It included many new requirements and processes, including building activities and work repairing ships in harbour. Further amendments and alterations were made in 1948 and 1959, with a consolidating measure in 1961 giving rise to today's Act. However, much of the Act has been repealed by the Health and Safety at Work etc. Act 1974 and replaced with generally-applicable Regulations or Codes of Practice. It will have been repealed in its entirety by the end of 1996.

Until the end of 1996, the Factories Act 1961 will continue to apply to premises within its scope and equipment within them which were first taken into use before 1st January 1993. For a summary of the situation, see the Sections in this Module covering the Workplace (Health, Safety and Welfare) Regulations and the Provision and Use of Work Equipment Regulations.

The Act applies only to premises defined within it, (notably factories, of course) and not to other work places. This was one of the difficulties overcome by the Health and Safety at Work Act and later EC-driven Regulations, with their emphasis on obligations based more on relationships than on narrow definitions of types of premises, and removing the need to interpret what is meant by "a factory" for new premises.

OBJECTIVES OF THE FACTORIES ACT

This Act contains many specific and detailed requirements. In the "factory" environment, there are four potential sources of injury to be addressed. These are: the premises themselves, physical or hardware elements which are part of or arise from the activities carried out there, hygiene and welfare aspects, and the workforce. The Act deals with these, although in a much less comprehensive way and ignoring significant aspects of modern practice.

The Act is divided into fourteen parts. There now follows an overview of these, but in order to appreciate the application of the legislation it would be necessary to study the text itself, using one of the annotated commentaries such as Redgrave's Health and Safety in Factories.

HOW THE OBJECTIVES ARE MET

Part I deals with the **health** general provisions, and predominantly those causal elements of accidents associated with the premises. Every factory must be kept clean and tidy, free from smells from drains and sanitary conveniences. Dirt and refuse, including discarded process material, must be removed daily and floors must be cleaned weekly. Walls, partitions and ceilings must be washed down every 14 months, and where painted must be repainted every 7 years (**Section 1**). Factory premises must not be overcrowded (**Section 2**) with workrooms maintained at an adequate temperature. Combustion products of the heating process must not enter the work place (**Section 3**). Adequate fresh air ventilation is required in each work room (**Section 4**). Effective provision must be made for sufficient and suitable lighting in every part of the factory in which persons are working or

The Factories Act 1961

passing through (**Section 5**). Adequate floor drainage is required where "wet" processes are carried on (**Section 6**), and adequate sanitary conveniences with lighting must be provided, maintained and kept clean (**Section 7**). **Section 10A** relates to the medical examination of workers, which is to be permitted if required by the Employment Medical Advisory Service.

Part II deals with **safety** general provisions as they relate to the physical dangers associated with factory processes and work. **Sections 12-14** deal with the fencing or guarding of dangerous parts of machinery. Every flywheel and moving part of a "prime mover" (this term includes an engine, a motor or generator) must be securely fenced. This includes parts of electric generators and rotary converters, and their flywheels, unless they are safe by position (**Section 12**). However, where practicable this exception regarding electric plant should be ignored and secure fencing provided. Every part of "transmission machinery" (that is, shafts, wheels, drums, pulleys, driving belts, clutches – any object that passes energy along by means of rotation or passage of a rope or belt) must be securely fenced unless it is safe to everyone by virtue of its position. Power cut-offs must be fitted to transmission machinery (**Section 13**).

Dangerous parts of all machinery other than prime movers and transmission machinery must be fenced, whether they are power-operated or not (**Section 14**). This Section is the most famous provision of the Factories Act, and is often quoted as the best example of an absolute duty or strict liability to do something regardless of cost or other consequences.

Section 15 makes provisions for specially-trained and certificated people, called "machinery attendants" to approach unfenced machinery under strict conditions. All fencing provided must be of substantial construction, maintained and repaired and kept in position when the machine is in use (**Section 16**). Duties are placed on suppliers of machinery to ensure it is appropriately guarded (**Section 17**). Vessels, pits etc. containing scalding, corrosive or poisonous liquids must have adequate and suitable edge protection. Access ways around such areas must be of an adequate width and provided with adequate edge protection (**Section 18**). Trapping areas between the moving elements of self-acting machines and fixed structures must be prevented (**Section 19**).

Women and young people (that is, under 18 years) are not allowed to clean dangerous parts of machinery whilst in motion or at rest (**Section 20**). Young persons are not allowed to work at certain machines, such as milling machines, unless they have been adequately

trained or are under the direct supervision of a competent person (**Section 21**).

Hoists and lifts must be of good construction, thoroughly examined (every 6 months), marked with their safe working load (SWL) and properly maintained. Hoistways and liftways must be enclosed and fitted with interlocking gates (**Section 22**). Additional requirements for hoists and lifts carrying people include automatic over-run devices, cut-outs to prevent operation of the lift or hoist whilst the gate is open, and dual suspension ropes (**Section 23**). Door openings used during hoisting or lowering of materials, such as teagle openings, must be securely fenced with adequate handholds provided. Fencing, usually chains, must be properly maintained and kept in position except when raising or lowering (**Section 24**). Chains, ropes, slings etc. must be of good construction, sound material, adequate strength and free from defects. They must be tested and examined before first use (except for fibre rope and fibre rope slings), examined every 6 months, and marked with their SWL. For multiple-leg chains or ropes etc. the SWL at different angles must be displayed on the premises. Chains or slings of 14mm bar or smaller, and all chains used in connection with molten metal or slag must be annealed once every 6 months, and all other chains and lifting tackle every 14 months (unless they could be damaged by this or are not regularly used) (**Section 26**). (There are considerable exemptions to this heat treatment requirement, contained in Certificate of Exemption No.1).

Cranes and other lifting machines must be of good construction, made from sound material, thoroughly examined every 14 months, tested and thoroughly examined before first use, marked with the SWL(s). If a jib crane, it must be fitted with an automatic safe load indicator (ASLI) or load charts where the load or radii are variable. Rails and/or tracks must be of good construction. Steps must be taken to prevent persons working at raised levels being struck by overhead travelling cranes (**Section 27**).

All access ways such as floors and stairs must be of good construction and properly maintained. They must be kept free from obstruction and non-slippery, and steps are required to have handrails as detailed in the Section. Openings in floors are to be fenced except where this is impracticable. Ladders are to be of adequate construction and properly maintained (**Section 28**).

A safe means of access and a safe place of employment must be provided, including the provision of adequate edge protection or its equivalent in circumstances where a person can fall more than 2m. (**Section 29**).

No person is allowed to enter or remain in a confined space where dangerous fumes are liable to be present unless the steps specified have been taken to make entry safe (for example the provision of suitable breathing apparatus) (**Section 30**).

Section 31 details the precautions required to prevent explosions and fires resulting from dusts, gases, vapours or substances. **Sections 32 to 38** refer to requirements now covered by the Pressure Systems and Transportable Gas Containers Regulations 1989, and are being progressively revoked, except for **Section 34** which prohibits entry into linked steam boilers unless specified precautions are taken.

Every gas holder with capacity greater than 140 m^3 must be of sound construction, properly maintained, examined and marked as directed (**Section 39**).

Part III deals with **welfare general provisions**. Adequate supplies of drinking water must be provided at convenient points (**Section 57**). Washing facilities must be adequate and suitable, including hot and cold running water, soap and towels. Facilities must be accessible and kept clean (**Section 58**). Accommodation for non-work clothing must be provided, with drying arrangements as specified (**Section 59**). **Section 60** refers to seating facilities which are to be provided.

Part IV covers **special provisions**. **Section 68** deals with humid factories, and **Section 69** with underground rooms.

Part V is repealed. **Part VI** details provisions regarding employment of women and young persons, and its contents should be examined as required. Further Parts deal with special applications of the Act, for example to docks, construction sites (**Part VII**), homeworkers (**Part VIII**), notices, returns and records (**Part X**) which also places duties on employees, administration (**Part XI**), and offences, penalties and legal proceedings (**Part XII**). The application of the Act and its interpretation is important, and is covered in **Parts XIII and XIV** respectively.

WHAT IS A FACTORY?

The definition of "factory" can be found in **Section 175(1)**, and is extensive. However there have been many court cases over the years which have turned upon whether duties were owed, in turn a question of the exact meaning of the words in the definition. Many other kinds of premises are technically factories even though they do not fit the definition, because they are referred to in **Section 175(2)**. The main features of the definition of a "factory" are that:

a) there must be premises involved, the boundaries of which can be defined

b) within those premises there is manual labour going on in connection with one of the processes specified

c) the processes or purposes must involve the making, altering, repairing, ornamenting, finishing, cleaning, washing, breaking up, demolition or adapting for sale of any article or part of an article – or the slaughtering or confining or animals which will be slaughtered (not being a cattle market)

d) the processes must be carried on by way of trade or for gain (this includes local authority and Crown premises as "gain" does not necessarily require profits to be made).

Examples of "articles" include any commodity regardless of its form – water and coal gas are articles, live animals are not, according to decided cases. "Adapting for sale" is a matter of degree. The article has to be made in some way different to what it was before. Water pumping stations are not factories under this test, as they only distribute the article, but filtration plants are factories because they adapt water for sale.

REVISION

The Factories Act is divided into **fourteen** parts:

Part I   – Health (general provisions)
Part II   – Safety (general provisions)
Part III   – Welfare (general provisions)
Part IV   – Special provisions to the above
Part V   – (revoked)
Part VI   – Employment of women and young persons
Part VII   – Special applications and extensions of the Act

Part VIII – Homeworkers
Part IX   – Wages
Part X   – Notices, records, returns & duties of employees
Part XI   – Administration
Part XII   – Offences, penalties & legal proceedings
Part XIII – Application of the Act
Part XIV – Interpretation.

The Offices, Shops and Railway Premises Act 1963

INTRODUCTION

Following the success of the Factories Act in bringing improvements to working conditions it was realised that improvements should be sought in other areas. Because of the specific nature of the Factories Act, only the health and safety of those persons employed in "factory" premises was covered. The need for similar specific legislation dealing particularly with offices and shops was satisfied with the implementation of the Offices, Shops and Railway Premises Act 1963. The Act received its Royal Assent in July 1963, but much of its provisions were not implemented until some twelve months later.

Eleven years later, with the implementation of the Health and Safety at Work etc. Act the general health and safety of all persons at work was protected, including those in the premises covered by this Act. However, much of the specific legislation contained in the Offices, Shops and Railway Premises Act is still in force for those specific premises covered by it. It will have been repealed in its entirety by the end of 1996.

Until then, the Offices, Shops and Railway Premises Act 1963 will continue to apply to premises within its scope and equipment within them which were first taken into use before 1st January 1993. For a summary of the situation, see the Sections in this Module covering the Workplace (Health, Safety and Welfare) Regulations and the Provision and Use of Work Equipment Regulations.

The health, safety and welfare provisions of the Act follow closely the provisions identified in the Factories Act, such as machine guarding, ventilation, cleanliness and provision of sanitary conveniences, and in laying down general standards.

OBJECTIVES OF THE ACT

The main objective of the Act is to set standards of health, safety and welfare for employees in offices, shops and railway premises. As with the Factories Act, the requirements of this Act address the common areas of danger to the employee. These can be broadly categorised into three distinct areas – dangers from the working environment (the premises or buildings themselves), welfare provisions, and danger from hardware elements such as machinery in use.

WHAT IS AN OFFICE, SHOP OR RAILWAY PREMISE?

An **office** is any building or part of a building used for administration, handling money, operating telephones, all forms of clerical work (such as writing, filing, typing) and similar activity. Any part of a building which is an office is covered by the Act, even if the rest of the building is not (offices in clubs, schools or hospitals, for example). The term "building" extends to cover a hut or a shed if used as office premises.

A **shop** includes places where retail or wholesale trade is carried on. The definition extends to cover premises which sell meals or refreshments to the public for immediate consumption, (such as public houses) but private residential hotels and boarding houses are excluded except where they have offices, bars or restaurants open to the public. Canteens for staff working in premises covered by the Act are included as are launderettes and dry cleaning premises.

A **railway premise** is defined as any railway building situated in the immediate vicinity of the permanent way of stations, marshalling yards and signal boxes. Exclusions include railway running sheds (these are covered by the provisions of the Factories Act) and hostels for railway staff.

Premises covered by the provisions of the Act are identified in **Section 1**. Specific exclusions include premises where only self-employed persons work; premises where only immediate relatives of the employer are employed; outworkers' dwellings; premises where the sum of all the hours worked is normally not more than 21 each week; premises where only servicemen are employed; dockside wholesale fish premises; and parts of mines below ground.

HOW THE OBJECTIVES ARE MET

The occupier of the premises is responsible for compliance with the Act, but in some cases where the premises are held on a lease and do not constitute an entire building, or where there is multi-ownership of premises within a building, the owner is responsible **(Sections 42 and 43)**.

Those Sections dealing with the dangers from the *working environment* include:-

Cleanliness – all premises, furniture, fittings and furnishings must be kept clean. Floors and steps must be cleaned not less than once a week by washing, sweeping or some other suitable method (this does not apply to fuel storage premises in the open air). No refuse or dirt should be allowed to accumulate **(Section 4)**.

Overcrowding – the number of people employed must not cause a risk of physical injury or risk to health. At least 3.7 m^2 are required for each person who is

The Offices, Shops and Railway Premises Act 1963

habitually employed in the room and 11 m³ breathing space **(Section 5)**. In making these calculations furniture, fittings, machinery etc. is not taken into account, but it is a major consideration in determining whether a room is overcrowded within the meaning of **Section 5(1)**. The numerical space standards do not apply to a room to which members of the public are invited, for example most parts of a shop, although such rooms will still be subject to the general prohibition of unhealthy overcrowding.

Temperature – a reasonable temperature must be maintained, and except where a substantial proportion of the work involves severe physical effort, this temperature is defined as 16 degrees Celsius after the first hour **(Section 5)**. This minimum standard does not apply to public rooms where its maintenance is not reasonably practicable, or in circumstances where deterioration of goods is likely to occur as a result. In these cases, employees must have access to sources of warmth and an opportunity, given by the employer, to warm themselves. No harmful fumes from any heating appliance provided should be allowed to enter the workplace. A thermometer must be provided on each floor, conspicuously placed so that employees can check the temperature. A maximum temperature is not defined, but the need to provide a safe and healthy working environment applies (Section 2, Health and Safety at Work Act) and so reasonably practicable steps must be taken to keep the temperature at an appropriate level.

Ventilation – effective and suitable ventilation must be supplied in all workrooms. This can be either fresh or artificially purified air **(Section 7)**.

Lighting – suitable and sufficient lighting is required, either natural or artificial, in each workroom and access way. Windows or skylights must be kept clean as far as is reasonably practicable, but can be shaded or whitewashed to reduce heat or glare. Where artificial lighting is provided it must be properly maintained **(Section 8)**.

Movement – provision is made to minimise the risk of falls. Floors, passages, stairs, steps and gangways must be kept free from obstruction and properly maintained. Floor openings must be fenced. Handrails are required to stairs with open side(s) and precautions must be taken to prevent people from falling through the open side of any staircase **(Section 16)**.

The provisions made regarding *welfare provisions* include:–

Toilets – must be suitable and sufficient, kept clean and tidy with effective lighting and ventilation. Facilities must

be easily accessible. Arrangements can be made for employees of different employers to share facilities provided that the requirement of the Act is satisfied **(Section 9).** Regulations made under the Act specify more detailed requirements. These are the Sanitary Conveniences Regulations 1964 and became effective on 1st January 1966.

Washing Facilities – these must be suitable and sufficient including a supply of clean, hot and cold (or warm) running water, soap and clean towels (or another equivalent means of drying) **(Section 10)**. More specific requirements are detailed in the Washing Facilities Regulations 1964, which became effective on 1st January 1966, and are binding under the main Act.

Drinking Water – wholesome drinking water must be provided along with drinking cups (unless water is available in a fountain jet). Facilities must be readily accessible **(Section 11)**.

Seating – in circumstances where employees have an opportunity to sit, without detriment to their work, adequate seating must be provided **(Section 13)**.

Seating for sedentary work – seats, provided for people who normally perform their work seated, must be suitable (in design, construction and dimensions) for the worker and for the kind of work being performed. A footrest must be provided unless both feet can be supported adequately. Seats and footrests must be adequately supported when in use **(Section 14)**.

Eating Facilities – in circumstances where persons employed in shops eat meals on the premises, adequate facilities must be provided **(Section 15)**.

Accommodation for clothing – arrangements must be made to hang clothing. Where reasonably practicable drying facilities for clothing must be provided **(Section 12)**.

First-aid provisions are now made under the Health and Safety (First-Aid) Regulations 1981 (see Section 7(6)).

Fire provisions are now made under the Fire Precautions Act 1971 and associated Regulations (see Section 7(8)).

Those sections dealing with the dangers from *hardware elements* include:–

Machinery – dangerous parts of machinery must be fenced or otherwise made safe. All fencing provided must be substantially constructed, properly maintained and kept in position while dangerous parts are in motion **(Section 17)**. Persons under the age of eighteen years

The Offices, Shops and Railway Premises Act 1963

are not permitted to clean any machinery if it exposes them to a risk of injury from a moving part **(Section 18)**. No person is permitted to work at any machine which has been specified as "dangerous" by the Secretary of State under the Prescribed Dangerous Machines Order 1964 (such as slicing machines, dough mixers and guillotines) unless they have been fully trained or is under the direct supervision of an experienced person **(Section 19)**. It is also important to remember the general training requirements of the Health and Safety at Work Act and the Management of Health and Safety at Work Regulations 1992.

Miscellaneous

Premises must be notified to the relevant enforcing authorities on the prescribed Form (OSR1) **(Section 49)**. **Section 27** makes it an offence for an employee to misuse or interfere with the facilities provided to meet the requirements of the Act. Additional Sections deal with enforcement and penalties.

Revision

The OSRP Act makes provision for:

- Cleanliness
- Preventing overcrowding
- Temperature
- Ventilation
- Lighting
- Safe movement

- Toilets
- Washing facilities
- Drinking water
- Seating
- Accommodation for clothing
- Eating facilities
- Machinery safeguards.

The Fire Precautions Act 1971

INTRODUCTION

Fire legislation, like other health and safety legislation, has evolved over a period of time, having its roots in the Industrial Revolution. It developed gradually meeting individual and specific needs, as and when it was necessary, embracing diverse premises. This approach to its development led to a fragmented body of legislation not just in relation to its requirements but also its enforcement. This lack of uniformity created pressure for rationalisation and consolidation.

Reform commenced with the Report of the Holroyd Committee on the fire service in 1970. Members of this Committee recommended that the fire safety laws be consolidated into two distinct areas. The first dealt with fire precautions for new and altered premises (generally controlled under the Building Regulations) and the second dealt with premises already occupied. Control of this latter area was completed by the **Fire Precautions Act 1971**, which was the first Act devoted specifically to fire. At the time of writing, the Home Office is reviewing its proposals for Regulations to implement the fire requirements of the Workplace Directive, which will undoubtedly alter some if not all of this Act.

With the enactment of the Health and Safety at Work etc. Act 1974, a framework was constructed whereby general fire precautions in industrial or commercial premises were subject to one comprehensive code (generally under the control of the fire authorities) while the fire precautions relating to dangerous materials and processes came under another (controlled by the HSE).

It is important to remember that although a process of consolidation has taken place, fire precautions are still subject to complex legislation. This Section is concerned only with industrial and commercial premises to which the Fire Precautions Act applies. Other relevant pieces of legislation, which are not discussed here, are the Petroleum Consolidation Act 1928, and the Fire Services Act 1947. The Fire Safety and Safety of Places of Sport Act 1987 is briefly outlined below.

OBJECTIVES OF THE ACT

The main objective of the Act is to provide a framework for the control of fire in premises which may pose a risk. The control process is one of certification, placing exacting standards on the various premises.

HOW THE OBJECTIVES ARE MET

The requirements of the Fire Precautions Act were brought in over a period of time, its greater part being implemented by 1972. Essentially, an "enabling" Act, it applies to all premises actually in use, whether industrial, commercial or public. Its impact has been increased by successive pieces of legislation such as the Fire Precautions (Factories, Offices Shops and Railway Premises) Order 1989, the Fire Precautions (Hotels and Boarding Houses) Order 1972, the Fire Precautions (Special Premises) Regulations, and the Fire Precautions (Non-Certified Factories, Offices, Shops and Railway Premises) (Revocations) Regulations 1989. The Fire Precautions Act imposes requirements on the occupiers whether they are employers or not.

The application of the provisions of the Act turns on the central question of whether or not particular premises require a fire certificate.

● Under the Act "**designated use**" premises require a fire certificate from the local authority in order to operate **(Section 1(1))**. For premises to become "designated" the Secretary of State has to make an Order under the Act. He can only designate premises which fall in to one or more of the following classes:-

a) premises used as, or providing, sleeping accommodation;

b) premises used as an institute providing treatment or care

c) premises used for entertainment, recreation or instruction or any club, association or society;

d) premises used for teaching, training or research;

e) premises where public access is involved;

f) premises used as a place of work.

For places of work, the criteria requiring them to have a fire certificate have been laid out in the Fire Precautions (Factories, Offices, Shops and Railway Premises) Order 1989. There are four broad categories of premises detailed. They are:

a) premises where the aggregate number of employees employed at any one time is more than 20 on one floor

b) premises where the aggregate number of employees is more than 10 if they are employed on different floors

c) instances where two or more premises are linked (where the totals given in a) and b) above apply, even though more than one employer is involved) and

The Fire Precautions Act 1971

d) factory premises storing or using explosive or highly flammable materials (in these circumstances the fire authority may determine that there is no risk present and a fire certificate will not be issued. This is at the discretion of the fire authority **only**).

The Fire Safety and Safety of Places of Sport Act 1987 amends the Fire Precautions Act to allow a fire authority to grant exemptions from certification for "designated use" premises at any time if the premises present a "low risk". The fire authority normally has to have inspected the premises within the previous 12 months, unless the exemption is given on a first application. The exemption certificate must specify the maximum safe number of people on the premises at any one time, and can be withdrawn at any time (amended **Section 5 (A)**).

Some premises, where hazardous materials, chemicals or processes are involved, are termed "**special premises**" and are under the administrative control of the HSE. Detailed Regulations have been implemented to certificate the use of such premises with regard to fire precautions. These are the Fire Certificate (Special Premises) Regulations 1976 – under which fifteen types of premises are identified as requiring a fire certificate. The HSE is responsible for issuing the certificates under these Regulations, and the administrative control.

FIRE CERTIFICATES

A fire certificate **will** specify:- the use of the premises; the means of escape in case of fire; the means for securing that the means of escape can be used safely at all relevant times; the means for fighting fire, for example the number and type of fire extinguishers; the means for giving warnings of fire, such as fire alarms; and particular requirements for any explosives or highly flammable materials stored or used on the premises **(Section 6)**.

In addition, a fire certificate **may** specify the required maintenance of the fire escape including safe access to and along it; means of other fire precautions; training of employees and the keeping of training records; limitation of the number of people allowed on the premises at any one time and any other additional relevant fire precautions **(Section 6(2))**.

Where alterations are made to the structure of the building or to the working conditions carried on within that building, application for a revised fire certificate must be made to the relevant enforcing authority **(Section 8)**.

Premises which have been exempted from the requirement to have a fire certificate must be provided with an adequate means of escape in case of fire and a means of fighting fire **(Section 9)**.

SELF-ASSESSMENT QUESTIONS

1. Does your own workplace require a fire certificate? Can you say why? Examine it carefully, and see how the certificate is worded and what (if any) special requirements have been added by the local fire authority.

2. List the contents of a fire certificate. Why are they crucial to planning for fire safety in the workplace?

REVISION

A fire certificate is required when:-

- more than 20 people are employed on one floor
- more than 10 people are employed on different floors

- adjoining premises accumulate these numbers
- flammable or explosive substances are stored or used in premises
- special premises as defined by Orders.

The Control of Substances Hazardous to Health Regulations 1988 (COSHH)

INTRODUCTION

People in the work place are often exposed to substances which have the potential to damage their health. Many of these are present as a direct result of their use in a manufacturing process, some are formed by the process itself. Others are used in maintenance activities such as cleaning. Some hazardous substances occur naturally, such as microbiological agents, which can cause diseases like leptospirosis (Weil's Disease). The move to control all such substances came as a result of EC Directive 80/1107/EEC on "the protection of workers from the risk related to exposure to chemical, physical and biological agents at work". These Regulations cover all people at work, from quarries to offices, and apply to virtually all substances hazardous to health. They became fully effective on 1st. January 1990.

OBJECTIVES OF THE REGULATIONS

The main objective of the Regulations is to prevent workplace disease resulting from exposure to hazardous substances. Basic occupational hygiene principles are followed by the Regulations which introduce a control framework by requiring an adequate **assessment** of the risks to health arising from work activities associated with hazardous substances, the introduction of adequate **control** measures, **maintenance** of the measures and equipment associated with them, and **monitoring** the effectiveness of the measures and the health of employees.

WHAT IS A SUBSTANCE "HAZARDOUS TO HEALTH"?

This term includes any material, mixture or compound used at work or arising from work activities, which is harmful to people's health in the form in which it occurs in the work activity. Categories specifically mentioned are a) substances labelled as dangerous (toxic, very toxic, harmful, corrosive and irritant) under other statutory provisions, b) substances assigned an OES or MEL, c) harmful micro-organisms, d) substantial quantities of dust, and e) any substance creating a comparable hazard. A "substantial" concentration of dust is more than 10 mg/m³ (8-hour TWA) of total inhalable dust, and more than 5 mg/m³ similarly of total respirable dust, where no lower value is given for the substance.

Module B contains more details on the terms above, and those used in the Regulations, which will not be given a further explanation here.

HOW THE OBJECTIVES ARE MET BY THE REGULATIONS

Duties are imposed upon employers for the protection of employees who may be exposed to substances hazardous to health at work, and of other persons who may be affected by such work. Specific duties are also imposed upon the self-employed and employees.

An employer must not carry on any work which is liable to expose any employee to a substance hazardous to health unless a suitable and sufficient assessment has been made of the risks to health created by the substance at work and about the measures necessary to control exposure to it (**Regulation 6**).

This is an essential requirement; the assessment is a systematic review of the use of the substance present – its form and quantity, possible harmful effects, how it is stored, handled, used and transported as appropriate, the people who may be affected and for how long, and the control steps which are appropriate. The assessment should take all of these matters into account, taking care to describe actual conditions when assessing risk. The assessment should be written down for ease of communication. A later amendment requires all risk assessments to be reviewed every five years, in addition to the requirement to review them if they are suspected of being invalid.

The employer must ensure that the exposure of employees to substances hazardous to health is either prevented, or where this is not reasonably practicable, is adequately controlled (**Regulation 7**). This applies whether the substance is hazardous through inhalation, ingestion, absorption through the skin or contact with the skin. Prevention of exposure must be attempted in the first place by means other than the provision and use of personal protective equipment. **Regulation 7(2)** only permits use of personal protective equipment (PPE) where other control is not reasonably practicable. Any respiratory protective equipment supplied must comply with current requirements as to suitability (**Regulation 7(6)**). Prevention of exposure can be achieved by removing the hazardous substance, substituting it with a less hazardous one or using it in a less hazardous form, enclosing the process, isolating the worker(s), using partial enclosure and extraction equipment, general ventilation, and the use of safe systems of work.

Occupational exposure limits have been set (and are revised and updated yearly) for a wide range of substances. Where an MEL has been assigned, the level of exposure must be reduced as far as is reasonably practicable (**Regulation 7(4)**). The MEL should not be

The Control of Substances Hazardous to Health Regulations 1988 (COSHH)

exceeded. For substances assigned an OES, it is sufficient to reduce the level of exposure to that standard.

Employers providing control measures must ensure they are properly used, and every employee must make full and proper use of what is provided, reporting any defects to the employer and doing their best to return it after use to any accommodation provided for it (**Regulation 8**).

Regulation 9 requires those control measures which are provided to be maintained in efficient working order and good repair. If the controls are engineering controls, they should be examined and tested at suitable intervals. For local exhaust ventilation equipment, this is defined as at least once in every 14 months generally. The Regulation also requires examination and testing of respiratory protective equipment at suitable intervals, where the equipment is provided under Regulation 7. Records of these tests and examinations have to be kept, and a summary of them has to be kept for at least 5 years.

Monitoring of exposure must be carried out (**Regulation 10**) where it is necessary to ensure that exposure is adequately controlled. For listed processes and substances, the frequency is specified. Records of this monitoring have to be kept for at least 5 years, or 30 years where employees can be personally identified. Monitoring would be required where: there could be serious risks to health if control measures should deteriorate; it cannot be guaranteed that exposure limits are not being exceeded, or that control measures are working properly.

Health surveillance is required where appropriate, which would be where there is significant exposure to a substance listed in Schedule 5 to the Regulations; or − where there is an identifiable disease or adverse health effect which may be related to the exposure, and there is a reasonable likelihood that this may occur under the particular work conditions, and there are valid techniques for detecting indications of it (**Regulation 11**). The surveillance can be done by doctors, nurses or trained supervisors making simple checks. Where the surveillance is carried out, records or copies of them must be kept for 40 years from the date of the last entry.

Where any of his employees are or may be exposed to substances hazardous to health, the employer must provide them with suitable and sufficient information, instruction and training for them to know the health risks created by the exposure and the precautions which should be taken (**Regulation 12**). The information must include the results of any monitoring and the collective (non-personalised) results of health surveillance.

Provision is made in **Regulation 13** for the control of certain fumigation operations, generally requiring notices to be posted beforehand.

These Regulations repealed a large number of existing statutes covering specific industries, including the Chemical Works Regulations 1922, the Hides and Skins Regulations 1921 and the Chromium Plating Regulations 1931. The Ministry of Defence is exempt from certain of the requirements of the Regulations, and there are a number of exceptions to them which are mostly covered in **Regulation 3**. Notably, health surveillance requirements are not extended to non-employees, but the training, information and monitoring requirements of **Regulations 10, 12(1) and 12(2)** are extended to them if they are on the employer's premises. Regulations **10 and 11** do not apply to the self-employed, but all the others do. The Regulations do not extend to normal ship-board activities of the crew of a sea-going ship or its master.

Regulations 6 to 12 do not have effect in some circumstances, covered in **Regulation 5**. These are: where the Control of Lead at Work Regulations 1980 apply; where the Control of Asbestos at Work Regulations 1987 apply; where the hazardous substance is hazardous solely because of its radioactivity, flammability, explosive properties, high or low temperature or high pressure; where the health risk to a person derives because the substance is administered in the course of medical treatment; or below ground in a mine.

The Control of Substances Hazardous to Health Regulations 1988 (COSHH)

SELF-ASSESSMENT QUESTIONS

1. A new substance is to be introduced into your work place. What basic steps should be taken to ensure compliance with the COSHH Regulations?

2. Select an area of your work place and identify substances which are defined as hazardous to health. Attempt to find examples from the operational categories of process, raw materials, engineering, cleaning, service, byproduct.

REVISION

The Control of Substances Hazardous to Health Regulations (COSHH) require:

- Assessment of health risks and selection of suitable control measures
- Practical control of the risks

- Maintenance of control measures
- Monitoring of exposure of employees, including health surveillance
- Monitoring of effectiveness of controls
- Information, instruction and training of the work-force.

The Electricity at Work Regulations 1989

INTRODUCTION

The dangers of the use of electricity have been discussed previously in Module A. Legislative control of electrical matters in the past has been concerned with not just the fundamental principles of electrical safety but also specific and detailed requirements relating to particular plant and activities. Until recently, the Electricity (Factories Act) Special Regulations 1908 and 1944 have controlled the use of electricity at work. These Regulations were enforced under the Factories Act 1961, and continued under the Health and Safety at Work Act 1974. The limiting factor was that the Regulations were only applicable to "factory" premises as defined (see Section 5), and thus applied to a limited number of people and premises. Also, some of the requirements of the Regulations had become outdated as technology advanced and working and engineering practices changed to accommodate it.

All the provisions of the Provision and Use of Work Equipment Regulations 1992 are relevant to these Regulations, which came into force on 1st April 1990. They removed the above drawbacks and provide a coherent and practical code which applies to all work areas and all workers.

OBJECTIVES OF THE REGULATIONS

The Regulations introduce a control framework incorporating fundamental principles of electrical safety, applying to a wide range of plant, systems and work activities. They apply to all places of work, and electrical systems at all voltages.

HOW THE OBJECTIVES ARE MET BY THE REGULATIONS

The Regulations revoke the earlier Electricity Regulations, and generally consist of requirements which have regard to principles of use and practice, rather than identifying particular circumstances and conditions. Action is required to prevent danger and injury from electricity in all its forms. Employers (including managers of mines and quarries), self-employed people and employees all have duties of compliance with the regulations so far as they relate to matters within their control, and these are all known as **duty holders**. Additionally, employees are required to co-operate with their employer so far is necessary for the employer to comply with the Regulations (**Regulation 3**).

All electrical systems must be constructed and maintained at all times to prevent danger, so far as is reasonably practicable (**Regulation 4(1) and 4(2)**). An **electrical system** is a system in which all electrical equipment is, or may be, electrically connected to a common source of electrical energy, and includes the source and the equipment. **Electrical equipment** includes anything used, intended to be used or installed for use to generate, provide, transmit, transform, rectify/convert, conduct, distribute, control, store, measure or use electrical energy – a pretty comprehensive definition! **Danger** means the risk of injury – where "injury" in this context means death or personal injury from electric shock, burn, explosion or arcing, or from a fire or explosion initiated by electrical energy, where any such death or injury is associated with electrical equipment.

Every work activity (including operation, use and maintenance of and work near electrical systems) shall be carried out so as not to give rise to danger, so far as is reasonably practicable. (**Regulation 4(3)** – see also further requirements of **Regulations 12, 13, 14 and 16**).

Equipment provided for the purpose of protecting persons at work near electrical equipment must be suitable, properly maintained and used (**Regulation 4(4)**). The term "protective equipment" as used here has a wide application, but typically it includes special tools, protective clothing or insulating screening equipment, for example, which may be necessary to work safely on live electrical equipment.

No electrical equipment shall be put into use where its strength and capability may be exceeded, giving rise to danger (**Regulation 5**). This requires that before equipment is energised the characteristics of the system to which it is connected must be taken into account, including those characteristics under normal, transient and fault conditions. The effects to be considered include voltage stress and the heating and electromagnetic effects of current.

Electrical equipment must be protected and constructed against adverse or hazardous environments (such as mechanical damage; weather; temperature or pressure; natural hazards; wet, dirty or corrosive conditions; flammable or explosive atmospheres) (**Regulation 6**).

All conductors in a system which could give rise to danger must be insulated, protected or placed so as not to cause danger (**Regulation 7**). A **conductor** means a conductor of electrical energy. The danger to be protected against generally arises from differences in electrical potential (voltage) between circuit conductors and others in the system, such as conductors at earth potential. The conventional approach is to insulate them, or place them so that people cannot receive electric shocks or burns.

Precautions shall be taken, by earthing or by other suitable means, to prevent danger from a conductor (other than a circuit conductor) which may become charged, either as a result of the use of the system or of a fault in the system (**Regulation 8**). A **circuit conductor** means any conductor in a system which is intended to carry current in normal conditions, or to be energised in normal conditions. This definition does not include a conductor provided solely to perform a protective function, as for example an earth connection. The scope of Regulation 8 is therefore such that it includes as requiring to be earthed a substantial number of types of conductor, including combined neutral and earth conductors, and others which may become charged under fault conditions such as metal conduit and trunking, metal water pipes and building structures.

Restrictions are made on the placing of anything which might give rise to danger (fuses, for example) in any circuit conductor connected to earth (**Regulation 9**).

Connections used in the joining of electrical systems must be both mechanically and electrically suitable (**Regulation 10**). This means that all connections in circuits and protective conductors, including connections to terminals, plugs and sockets and any other means of joining or connecting conductors should be suitable for the purposes for which they are used. This applies equally to temporary and permanent connections.

Systems must be protected from any dangers arising from excess current (**Regulation 11**). The means of protection is likely to take the form of fuses or circuit breakers controlled by relays. Other means are also capable of achieving compliance.

There must be suitable means provided for "cutting off" energy supply to and the isolation of electrical equipment. These means must not be a source of electrical energy themselves (**Regulation 12(1) and (2)**). In circumstances where switching off and isolation is impracticable, as in large capacitors, for example, precautions must be taken to prevent danger so far as is reasonably practicable (**Regulation 12(3)**). Switching off can be achieved by direct manual operation or by indirect operation by "stop" buttons. Effective isolation includes ensuring that the supply remains switched off and that inadvertent reconnection is prevented. This

essential difference between "switching off" and "isolation" is crucial to understanding and complying with **Regulation 13**. This requires that adequate precautions are taken to prevent electrical equipment made dead to be worked on, from becoming electrically charged, for example by adequate isolation of the equipment.

No person shall be engaged in any work near a live conductor (unless insulated so as to prevent danger) unless a) it is unreasonable in all circumstances for it to be dead, b) it is reasonable in all circumstances for persons to be at work on or near it while it is live, and c) suitable precautions are taken to prevent injury (**Regulation 14**).

Adequate working space, means of access and lighting must be provided at all electrical equipment on which or near which work is being done which could give rise to danger (**Regulation 15**).

Restrictions are placed on who can work where technical knowledge or experience is necessary to prevent danger or injury (**Regulation 16**). They must possess the necessary knowledge or experience or be under appropriate supervision having regard to the nature of the work.

Regulations 17-27 inclusive give detailed additional requirements relating to mines and quarries, specifically those premises covered by the Mines and Quarries Act 1954.

Regulation 29 provides a means of defence in legal proceedings (See Section 2), where a person charged can show that all reasonable steps were taken and all due diligence was exercised to avoid committing the offence.

The requirements of the Regulations do not extend to the master or crew of a sea-going ship, or to their employer in relation to normal crew ship-board activities under the direction of the master, and they do not extend to any person in relation to an aircraft or hovercraft moving under its own power (**Regulation 32**).

The Electricity at Work Regulations 1989

SELF-ASSESSMENT QUESTIONS

1. What are the conditions under which live electrical working is permissible?

2. What are the qualities a person must have to carry out work with electrical apparatus?

REVISION

The Electricity at Work Regulations refer to:

- Construction and maintenance of electrical equipment
- Carrying out work activities near electrical systems
- Provision of protective equipment
- Putting electrical equipment into use
- Protection of electrical equipment
- Precautions required in relation to conductors
- Suitability of connections
- Protection from excess current
- Switching off and effective isolation of current
- Restriction of work on "live" conductors
- Provision of adequate space, access and lighting
- Restrictions on personnel to carry out electrical work.

INTRODUCTION

Although exposure to noise at work has been recognised as damaging to health for a number of years, until recently there has been little specific legal control. The Agriculture (Tractor Cabs) Regulations 1974, the Offshore Installations (Operational Safety, Health and Welfare) Regulations 1976 and the Woodworking Machines Regulations 1974 (Regulation 44 of which related to noise and is now revoked) have specific controls included in them, and there has been a general use of the Code of Practice issued in 1972 by the Department of Employment "Code of Practice for reducing the exposure of employed persons to noise" to determine compliance with the general duties imposed by the Health and Safety at Work Act.

From 1st. January 1990, the Noise at Work Regulations became effective, and introduced a framework for controlling exposure to workplace noise and ensuring compliance with the EC Directive on the protection of workers from risks relating to exposure to noise at work. The Regulations require the protection of all persons at work, with a broad general requirement for employers to reduce the risk to employees arising from noise exposure as far as is reasonably practicable (**Regulation 6**). The Regulations also affect the self-employed and trainees.

OBJECTIVES OF THE REGULATIONS

The Regulations introduce a control framework by requiring assessment of the extent of the problem, by carrying out noise surveys to identify work areas and employees at risk; control by engineering measures, isolation or segregation of affected employees, supply and use of protective equipment and administrative means; and monitoring by reassessment at intervals to ensure that the controls used remain effective.

HOW THE OBJECTIVES ARE MET BY THE REGULATIONS

Regulation 6 imposes a general duty to reduce the risk of hearing damage to the lowest level that is reasonably practicable.

Three noise **action levels** are defined, which determine the course of action an employer has to take if his employees are exposed to noise at or above the levels. These are:

- First Action Level – daily personal noise exposure ($L_{EP,d}$) of 85 dB(A)
- Second Action Level – daily personal noise exposure of 90 dB(A)
- Peak Action Level – peak sound pressure level of 200 Pascals (Pa) or more. (NB. This Level is important in circumstances where workers are subjected to small numbers of loud impulse noises during an otherwise relatively quiet day, for example during piling operations.)

Where the daily noise exposure exceeds the First Action Level, employers must:

- Carry out noise assessments (**Regulation 4**) and keep records of these until new ones are made. Competent persons only should carry out the assessments (**Regulation 5**)
- Provide adequate information, instruction and training for the employees, about the risks to hearing, the steps to be taken to minimise the risks, how employees can obtain hearing protectors if they are exposed to levels between 85 and 90 dB(A)$L_{EP,d}$ and their obligations under the Regulations (**Regulation 11**).
- Ensure that hearing protectors (complying with current requirements as to suitability) are provided to those employees who ask for them (**Regulation 8(1)**) and that the protectors are maintained or repaired as required (**Regulation 10(1)(b)**).
- Ensure so far as is practicable that all equipment provided under the Regulations is used (**Regulation 10(a)**) – apart from the hearing protectors provided on request as noted above.

Where the noise exposure exceeds the Second Action Level or the Peak Action Level, employers must in addition to the requirements detailed above:

- Take steps to reduce noise exposure so far as is reasonably practicable by means other than provision of hearing protection (**Regulation 7**). This is an important requirement, introducing good occupational hygiene practice to ensure hazards are designed out from a process as opposed to employee involvement only to reduce the risks.
- Establish hearing protection zones, marking them with notices so far as is reasonably practicable (**Regulation 9**).
- Supply hearing protection (complying with current requirements as to suitability) to those exposed (**Regulation 8(2)**) and ensure they are worn by them (**Regulation 10(1)(a)**), so far as is practicable.
- Ensure that all those entering marked hearing protection zones use hearing protection, so far as is reasonably practicable.

The Noise at Work Regulations 1989

Duties are imposed on workers by the Regulations to ensure their effectiveness. The requirements (**Regulation 10(2)**) are:

- On being exposed to noise at or above the First Action Level, they must use noise control equipment other than hearing protection which the employer may provide and report any defects discovered to the employer.
- If exposed to the Second Action Level or Peak Action Level, they must in addition wear the hearing protection supplied by the employer.

The Regulations also extend duties placed on designers, importers, suppliers and manufacturers under Section 6 of the Health and Safety at Work Act to provide information on the noise likely to be generated, should an article (which can mean a machine or other noise-producing device) produce noise levels reaching any of the three Action Levels (**Regulation 12**). In practice, this means supplying data from noise tests conducted on the machine, in addition to other information regarding safe use, installation etc. already required by Section 6.

SELF-ASSESSMENT QUESTIONS

1. The exposure of some of your workers has been measured and found to fall between the First and Second Action Levels. What steps should be taken to ensure compliance with the Noise at Work Regulations?

2. What are the main differences between the action required to be taken when the Second or Peak Action Levels are exceeded, and when the First Action Level is exceeded?

REVISION

The Noise at Work Regulations require:

- Adequate assessment
- Assessment records
- Reduction of the risk of hearing damage
- Reduction of noise exposure, starting with engineering controls
- Provision and maintenance of hearing protection
- Provision of information and training for employees
- Manufacturers and others to provide noise data.

INTRODUCTION

It is important to note that these Regulations do not apply to new equipment taken into use after 1st January 1993, which is subject to the Provision and Use of Work Equipment Regulations 1992. For this equipment, the Abrasive Wheels Regulations have been repealed except for the mandatory training requirements of Regulation 9 and the Schedule. Under transitional provisions, the Abrasive Wheels Regulations continue to apply to existing (pre-1993) equipment until 1st January 1997. These transitional arrangements also affect Regulation 17, in relation to the workplace (floors), which has been supplanted for premises newly taken into use by Regulation 12 of the Workplace (Health, Safety and Welfare) Regulations 1992.

The Abrasive Wheels Regulations became fully effective on 2nd. April 1972. There were a number of reasons for their introduction, not the least of which was the number of injuries arising from the use of abrasive wheels in industry. Statistics had shown that a high proportion of these were due to faulty mounting of the wheel – the marriage of an often heavy piece of abrasive compound to a part of a machine designed to rotate at high speed. Another major reason for the Regulations was to remove an existing legal anomaly.

Section 14 of the Factories Act 1961 contains the absolute duty to fence every dangerous part of a machine – even to the extent that the machine may, as a result, become unusable [John Summers (John) & Sons Ltd. v Frost [1955] AC 740]. This duty also applied to abrasive wheels, and required the wheel to be completely enclosed in order to satisfy the requirement. Unfortunately, the majority of abrasive wheels operate by being fixed in position and having metal work pieces brought up against the exposed wheel or section of wheel, which was therefore not legal. When the Abrasive Wheels Regulations became effective, the anomaly was removed as the section of wheel necessarily in use was allowed to remain unguarded. Regulation 3 specifically exempts abrasive wheel machines from compliance with Section 14 of the Factories Act (but the Regulations do contain stringent guarding requirements which must be complied with by abrasive wheel machines).

The Regulations apply to all factory premises, and premises defined as factories by the Factories Act, such as construction sites, where abrasive wheel machines are used. They do not apply when the machines are used elsewhere, although the Regulations can be used as a standard which can be achieved "reasonably practicably" under the Health and Safety at Work Act and should therefore be followed in all employment circumstances.

DEFINITION OF "ABRASIVE WHEEL"

Regulation 2 defines an abrasive wheel as:

- a wheel, cylinder, disc or cone consisting of abrasive particles held together by mineral, metallic or organic bonds, either natural or artificial;
- a mounted wheel or point, and a wheel or disc having separate segments of abrasive material;
- a wheel or disc made from metal, wood, cloth, felt, rubber or paper, having a surface (wholly or partly) consisting of an abrasive material;
- a wheel, disc or saw having a rim or segments consisting of diamond abrasive particles.

There is scope within the Regulations to make exceptions for certain classes of wheel from all the requirements (**Regulation 4**) and for issuing certificates of exemption from certain of the requirements (**Regulation 5**). These powers have been used in some cases, for example abrasive wheels used for the manufacture of crystal glass are exempted from **Regulations 9(1) and 16**.

OBJECTIVES OF THE REGULATIONS

Accidents involving these machines can cause extensive injuries, and are usually due to the bursting or disintegration of the wheel, or contact with the wheel by the operator. Regulations designed for protection must therefore contain control measures to do so.

Wheel bursts can be caused by overspeeding the wheel, incorrect mounting of the wheel on the spindle or in the chuck, the selection of a faulty wheel or one inappropriate for the job, misuse, or by damage caused by contact with a guard.

Contact with the wheel by the operator could result from the guard being removed or wrongly adjusted, from damaged operating controls, an untidy work area causing tripping or slipping, from an insecure work piece, or from clothing entanglement.

THE REGULATIONS

The Regulations meet their objectives by requiring the following controls:

- The maximum permitted speed of the wheel in r.p.m must be marked on the wheel or its washer. If wheels are under 55mm in diameter, they can be identified

The Abrasive Wheels Regulations 1970

by a mark so that the maximum speed can be checked on a notice displayed in the work area, stating the maximum permitted r.p.m.'s for wheels in use. (Exceptions have been made to this requirement.) Having indicated the maximum speed of use, no wheel shall be operated in excess of that (**Regulation 6**).

- Abrasive wheel machines must also have a notice fixed prominently on them, giving the maximum spindle speed or the working speeds if variable. The spindle speed must not exceed the maximum wheel working speed (**Regulation 7**). It is illegal to sell an abrasive wheel machine unless its spindle speed maximum r.p.m. is marked on it (**Regulation 19**).

These requirements are intended to eliminate wheel bursting by controlling the maximum speed of the wheel in relation to the maximum speed of the machine on which it is mounted – the former must always exceed the latter. Bursting is also controlled by the requirement to select the right wheel for the job (**Regulation 13**) – most wheels are marked according to the type of material they are suitable for cutting or grinding, but a voluntary British Standard marking code exists which may be followed by manufacturers in whole or part.

Training in wheel selection and the mounting technique appropriate to the type of wheel and machine is essential, and mandatory:

- Every wheel must be **properly mounted** on the machine (**Regulation 8**) – an absolute requirement.
- Every person who changes (mounts) an abrasive wheel must be trained, competent and appointed in writing to do so by his employer. The employer has to keep a Register of appointed persons, and give the employee a personal copy of the appointment notice. This does not apply to trainees under supervision, or to persons who change mounted wheels or points (**Regulation 9**). The training requirement is one of the few to be detailed within Regulations, and details are contained in a Schedule to the Regulations. It aims to ensure steps are taken to prevent bursting of wheels by ensuring correct mounting, to ensure correct selection by knowledge of the types of wheels appropriate for different jobs, and to prevent contact with the wheels by identifying and checking the protective measures of the machine. To achieve this, the Regulations distinguish the training requirements for different types of wheel. The Schedule to the Regulations gives the following training requirements:
- Provision of approved advisory literature (HSE Guidance Note PM32)
- Hazards arising from the use of abrasive wheels, and the precautions required

- Methods of marking abrasive wheels as to type and speed
- Methods of storing, handling and transporting abrasive wheels
- Methods of inspecting and testing abrasive wheels for damage
- The functions of all components; their correct use and assembly
- Dressing methods for abrasive wheels
- The correct adjustment of rests
- The requirements of the Regulations.

Accidents arising out of contact with the wheel, or its burst parts, are intended to be controlled by the Regulations covering guarding to be prevent contact or entanglement.

- Abrasive wheels must be guarded except where the use of guards makes this impracticable because of the nature of the work or the type of wheel used (**Regulation 10**). Some exemptions from this requirement have been made.
- Guards must be: capable of containing a wheel as far as is reasonably practicable if it fractures; properly maintained and secured; (designed to) enclose the whole of the wheel except for the part which needs to be exposed to do the work (**Regulation 11**).
- Conveniently-situated stopping/starting devices must be provided to all grinding machines.

Other contact-type accidents are to be prevented by:

- The requirement to ensure that floors in grinding areas are maintained in good condition, free from loose material and not slippery, as far as is reasonably practicable (**Regulation 17**).
- The securing of tool rests (where provided), and ensuring they are properly constructed, maintained and adjusted.

The final element of control places obligations on employees not to misuse or remove guards, flanges or rests, and to report any defects in the equipment to their employer (**Regulation 18**).

A copy of the Regulations must be displayed at the work place (Section 139, Factories Act 1961), and also a copy of the approved Cautionary Notice (**Regulation 16**). A further notice is required if mounted wheels or points are used.

OTHER RELEVANT LEGISLATION

It is important to remember that other statutory requirements will affect the use of abrasive wheels, and will impose other sets of obligations on management and workers alike. Examples of these are the Noise at Work Regulations 1989, Control of Substances Hazardous to Health (COSHH) Regulations 1988 (exposure to dusts), the Provision and Use of Work Equipment Regulations 1992, the Workplace (Health, Safety and Welfare) Regulations 1992, the Personal Protective Equipment at Work Regulations 1992, and the Management of Health and Safety at Work Regulations 1992.

REVISION

The Abrasive Wheels Regulations require:

- Display of maximum r.p.m. of wheel and spindle
- Correct wheel mounting
- Training for persons to mount wheels
- Records of training to be kept and notices displayed
- Provision of adequate guards and flanges

- Suitability
- Stopping and starting controls on machines
- Rests in good order where provided
- Maintenance of floors
- Employees to report defects, not to misuse equipment.

The Health and Safety (First-Aid) Regulations 1981

INTRODUCTION

The correct response to injured persons to give first-aid once an accident has occurred is of vital importance. It can mean the prevention of further injury, or even death. First-aid has two functions. First, it provides treatment for the purpose of preserving life and minimising the consequences of injury or illness until medical (doctor or nurse) help can be obtained. Second, it provides treatment of minor injuries which would otherwise receive no treatment, or which do not need the help of a medical practitioner or nurse. This definition of first-aid is included in **Regulation 2**. To ensure the appropriate steps have been taken to provide adequate facilities (trained personnel as well as equipment) in all places of work, the Health and Safety (First-Aid) Regulations were implemented on 1st. July 1982. They replace several pieces of legislation which imposed specific requirements for specific places of work, and set out rules covering all places of work.

OBJECTIVES OF THE REGULATIONS

The Regulations are supported by an Approved Code of Practice, and provide a framework for first-aid arrangements. This incorporates a flexibility, by setting objective standards to be achieved. Different types of premises, processes and industries can thus be covered by the same set of Regulations, requiring them to develop effective first-aid arrangements after making an assessment of the risks involved and the likely use of the facilities.

HOW THE REGULATIONS MEET THE OBJECTIVES

The Regulations meet their objectives by requiring that every employer must provide equipment and facilities which are adequate and appropriate in the circumstances for administering first-aid to his employees (**Regulation 3(1)**).

To ensure compliance with this requirement, an employer must make an assessment to determine the needs. Consideration of the following is required:

1. **Different work activities.** These need different provisions. Some, such as offices, have relatively few hazards and low levels of risk; others have more or more specific hazards (construction or chemical sites). First-aid requirements will be dependent on the type of work being carried out.

2. **Difficult access to treatment.** The provision of an equipped first-aid room may be required if ambulance access is difficult or likely to be delayed. The ambulance service should be informed in any case if the work is hazardous.

3. **Employees working away from employers' premises.** The nature of the work and its risk will need to be considered, and whether there is a work group or single employees.

4. **Employees of more than one employer working together.** Agreement can be made to share adequate facilities, with one employer responsible for their provision. Such agreement should be in writing, with steps taken by each employer to inform his employees of the arrangements.

5. **Provision for non-employees.** The Regulations do not require an employer to make first-aid provision for any person other than his employees, but liability issues and interpretation placed on the Health and Safety at Work Act may alter the situation, as, for example, in the case of a shop or other place where the public enter.

Having made this assessment, the employer will then be able to work out the number and size of first-aid boxes required. The Approved Code of Practice outlines minimum standards for their contents and facilities – at least one will always be required. Additional facilities such as a stretcher or first-aid room may also be appropriate.

The employer must ensure that adequate numbers of "suitable persons" are provided to administer first-aid. "Suitable persons" are those who have received training and acquired qualifications approved by the HSE, and any additional training which might be appropriate under the circumstances, such in relation to any special hazards (**Regulation 3(2)**) All relevant factors have to be taken into account when deciding how many "suitable persons" will be needed. These include:

1. **Situations where access to treatment is difficult.** First-aiders would be required where work activities are a long distance from accident and emergency facilities.

2. **Sharing first-aiders.** Arrangements can be made to share the expertise of personnel. Usually, as on a multi-contractor site, one contractor supplies the personnel.

3. **Employees regularly working away from the employer's premises.**

4. **The numbers of the employees**, including fluctuations caused by shift patterns. The more employees there are, the higher the probability of injury.

5. **Absence of first-aiders** through illness or annual leave.

In circumstances where the first-aider is absent – "in temporary and exceptional circumstances" – such as through sudden illness (but not through planned annual leave), an employer can appoint a person to take charge in an emergency and take charge of the equipment and facilities provided (**Regulation 3(3)**). Also, in appropriate circumstances, an employer can provide an "appointed person" instead of a first-aider. He must first consider the nature of the work, the number of employees and the location of the workplace (**Regulation 3(4)**). The "appointed person" is someone appointed by the employer to take charge of the situation (for example, to call an ambulance) if a serious injury occurs in the absence of a first-aider. It is recommended that the "appointed person" be able to administer emergency first-aid and be responsible for the equipment provided. In some situations, by virtue of the location, nature of work and the (small) number of employees, provision of an "appointed person" only will be adequate. However, **as a minimum** an employer must provide an "appointed person" at all times when employees are at work.

An employer must inform his employees about the first-aid arrangements, including the location of equipment, facilities and identification of trained personnel (**Regulation 4**). New employee induction training, and when commencing work in a new area are necessary times to do this. This is normally done by describing the arrangements in the safety policy statement, and the displaying of at least one notice giving details of the location of the facilities and trained personnel.

Self-employed people must ensure that adequate and suitable provision is made for administering first-aid while at work (**Regulation 5**). Again, an assessment has to be made of the likely hazards and risks to determine the extent and nature of what needs to be provided. It is also possible for the self-employed to make agreements with employers to share facilities.

These Regulations do not apply where the Diving Operations at Work Regulations 1981, the Merchant Shipping (Fishing Vessels) Regulations 1974 or the Merchant Shipping (Medical Scales) Regulations 1974 apply. The Regulations do not apply to vessels registered outside the UK, to mines (of coal, stratified ironstone, shale or fire clay), or to the armed forces (**Regulation 7**). **Regulations 8 and 9** make provision for mines not excluded by **Regulation 7** and for work offshore respectively. **Regulation 10** deals with repeals, revocations and modifications, the more notable of which include the repeal of Section 61 of the Factories Act 1961 and Section 24 of the Offices, Shops and Railway Premises Act 1963.

SELF-ASSESSMENT QUESTIONS

1. What considerations would have to be made to determine first-aid requirements for a construction site?

2. Explain the difference between a first-aider and an appointed person.

REVISION

The Health and Safety (First-Aid) Regulations require:

- The provision of adequate first-aid equipment

- The provision of adequately trained personnel
- Giving of information on first-aid provision to employees.

The Reporting of Injuries, Diseases and Dangerous Occurrences Regulations 1985 (Riddor)

INTRODUCTION

Information about the types of accidents which happen is a very useful tool to work with in the prevention of future events of a similar kind. The information gained can be used to indicate how and where problems occur, and demonstrates trends of time. It is important to distinguish between accidents, incidents and injuries – they are not the same. Injury can occur as a result of an incident: the injury and the incident together amount to an accident (the common term). Since 1985, when these Regulations replaced the Notification of Accidents and Dangerous Occurrences Regulations (NADO), the law has properly recognised in the title of the Regulations that it requires the collection of specified information about incidents which result in specified types of injury; in some cases it also requires information about specified incidents with the potential to cause serious physical injury, whether or not they produced such injury.

The enforcing authorities are interested in assembling such information because it gives them knowledge of trends and performance (failure) statistics. It also highlights areas for research, enforcement or future legislation.

These Regulations were implemented on 1st. April 1986.

OBJECTIVES OF THE REGULATIONS

The main purpose of the Regulations is to provide enforcing authorities with information on specific injuries, diseases and dangerous occurrences arising from work activities covered by the Health and Safety at Work Act. The authorities are able to investigate only a proportion of the total, so the Regulations aim to bring the most serious injuries to their attention quickly.

HOW THE REGULATIONS MEET THE OBJECTIVES

The Regulations meet their objectives by requiring the following:

• Where any person dies or suffers any of the injuries or conditions specified in Schedule 1, or where there is a "dangerous occurrence" as defined in Schedule 2, as a result of work activities, the "responsible person" must notify the relevant enforcing authority. This must be done by the quickest practicable means (usually the telephone) and send a report to them within 7 days (**Regulation 3(1)**). If the personal injury results in an absence of more than 3 calendar days, but does not fall into the categories specified as "major", the written report alone is required (**Regulation 3(3)**). The day of the accident is not counted when calculating absence.

• The "responsible person" may be the employer of the person injured, self-employed, someone in control of premises where work is being carried out, or someone who provides training for employment. The "responsible person" for reporting any particular injury or dangerous occurrence is determined by the circumstances, and the employment or other relationship of the person who is killed or suffers the injury or condition.

• The regulations cover employees, self-employed people and those who receive training for employment (as defined by the Health and Safety at Work Act), and also members of the public, pupils and students, hotel residents and other people who die or suffer injuries or conditions specified, as a result of work activity (**Regulation 2**).

• The "enforcing authority" is the body responsible for the enforcement of health and safety legislation relating to the premises where the injury or disease occurred. Usually, this will be the HSE or the local authority. In case of doubt, reports should be made to the HSE, which will forward the information to the correct enforcing authority if this is not the HSE.

Exceptions – The Regulations do not extend to cover: hospital patients who die or are injured whilst undergoing treatment in hospital, dental or medical surgeries; members of the armed services killed or injured whilst on duty; people killed or injured on the road (except where the injury or condition results out of: exposure to a substance conveyed by road, unloading or loading vehicles, maintenance and construction activities on public roads) (**Regulation 10**). They do not cover dangerous occurrences on public and private roads either.

Other exclusions from the reporting requirements (**Regulation 10(4)**) include:

• death or injury of anyone inside or in contact with civil or military aircraft
• death or injury of anyone employed on or carried by a merchant ship (unless they are shore-based and working on loading, unloading, repair or similar activities)
• a dangerous occurrence on board a merchant ship
• death or injury of anyone in connection with a statutory railway

The Reporting of Injuries, Diseases and Dangerous Occurrences Regulations 1985 (Riddor)

- death or injury of anyone where the Explosives Act 1875 applies
- an incident involving the escape of radioactive material
- agricultural poisoning.

Where death results within one year of a notifiable work accident or condition, the person's employer must notify the relevant enforcing authority in writing (**Regulation 4**). There is no prescribed Form for this purpose.

When reporting injuries and dangerous occurrences, the approved Form must be used (F2508). The reporting of diseases (on Form F2508A) specified in Appendix 1 is required only when the employer receives a written statement or other confirmation from a registered medical practitioner that the affected person is not only suffering from a listed disease but also that it has arisen in the manner also specified in the Appendix (**Regulation 5**).

Incidents involving the supply of flammable gas must be notified to the HSE and a written report must be sent on the approved Form (**Regulation 6**).

Records must be kept by employers and others of those injuries, diseases and dangerous occurrences which require reporting (**Regulation 7**). Records can be kept in the form of entries made in the Accident Book (Form BI510) with reportable injuries and occurrences clearly highlighted, by keeping photocopies of reports sent to the enforcing authorities, or on computer provided that they can be retrieved and printed out. Keeping of computer records of this type will require registration under the Data Protection Act if individuals can be identified.

Records should be kept at the place of work or business for at least three years from the date they were made. The enforcing authorities may request copies of such records which must then be provided (**Regulation 7**). Further information may be requested by the enforcing authorities about any injury, disease or dangerous occurrence (**Regulation 9**) – in some situations analysis of the initial reports might indicate patterns or categories indicating a need for further legislation or action, and a deeper study requiring more detailed information may then be carried by the HSE with the approval of the HSC.

There are additional provisions in **Regulation 8** for mines and quarries.

SELF-ASSESSMENT QUESTIONS

1. For your place of work, find out the enforcing authority, and the system followed to ensure that reportable injuries are properly notified.

2. Are the following reportable under the Regulations?

   a) A resident in a nursing home trips over a contractor's cable on the floor, breaking his leg.

   b) Two cars are involved in an accident on a motorway. One of the drivers, who is driving on business during normal working hours, is killed.

   c) An employee is unloading materials from a lorry parked in a road outside a building site. He is hit by a passing car, admitted to hospital for observation and released two days later.

REVISION

The Regulations require:

- Notification: of certain injuries, diseases and dangerous occurrences, as defined
- Record-keeping: of reports/notifications sent to the enforcing authority
- Further assistance: to be given to the enforcing authority which they may require.

Introduction to the Construction Regulations

The requirements of the Factories Act 1961 apply to "building operations" and "works of engineering construction" (i.e. construction work), but few of the Act's requirements actually deal with the safety of workmen in construction processes. Specific Regulations have therefore been made with the intention of protecting employees against the particular dangers arising in the industry. There are five sets of Regulations in all. They are:

Construction (General Provisions) Regulations 1961
Construction (Lifting Operations) Regulations 1961
Construction (Working Places) Regulations 1966
Construction (Health and Welfare) Regulations 1966
Construction (Head Protection) Regulations 1989

The Construction (Head Protection) Regulations 1989 were made under the Health and Safety at Work etc. Act 1974 with the earlier sets made under the Factories Act 1961.

Other Regulations apply to construction activities because they were also made under the Factories Act, and include:

Abrasive Wheels Regulations 1970
Highly Flammable Liquids and Liquefied Petroleum Gas Regulations 1972
Woodworking Machines Regulations 1974

Other "global" Regulations made under the Health and Safety at Work etc. Act 1974 apply to all workplaces, and thus to construction activities. These include:

Health and Safety (First-Aid) Regulations 1981
Control of Lead at Work Regulations 1980
Control of Asbestos at Work Regulations 1987
Control of Substances Hazardous to Health Regulations 1988
Electricity at Work Regulations 1989
Noise at Work Regulations 1989
Management of Health and Safety at Work Regulations 1992
Provision and Use of Work Equipment Regulations 1992
Manual Handling Operations Regulations 1992
Health and Safety (Display Screen Equipment) Regulations 1992

The requirements of these Regulations and those of the Health and Safety at Work etc. Act 1974, the Factories Act 1961 and the Offices, Shops and Railway Premises Act 1963 are dealt with in other Sections of this book. The Workplace (Health, Safety and Welfare) Regulations 1992 do not apply to construction work or to site offices, but amendments to the 1961 and 1966 Regulations are expected as a result of implementation of other Directives in due course.

APPLICATION OF THE REGULATIONS

'**Building operations**' and '**works of engineering construction**' are defined in Section 176(1) of the Factories Act 1961, and the major provisions of the Act are extended to those works by Section 127 and it is to these activities that the requirements of these Regulations are directed. In some circumstances, such as archaeological excavation, the work may not fall within the strict definitions so as to make the Construction Regulations applicable, but the work will still be subject to the requirements of the Health and Safety at Work etc. Act 1974. In these circumstances the specific Regulations will provide an indication of accepted standards to be achieved as far as is reasonably practicable. The terms are defined as follows:

'**Building operation**' means the construction, structural alteration, repair or maintenance of a building (including re-pointing, re-decoration and external cleaning of a structure). It also includes the demolition of a building and the preparation for, and laying the foundation of, an intended building. It does not include any operation which is defined as a "work of engineering construction".

'**Work of engineering construction**' means the construction of any railway line or siding (other than those on an existing railway). It also includes the construction, structural alteration or repair (including re-pointing and re-painting) or the demolition of any dock, harbour, inland navigation, tunnel, bridge, viaduct, waterworks, reservoir, pipe-line, aqueduct, sewer, sewage works or gasholder (except where carried on upon a railway or tramway). This definition has been extended to include any steel or reinforced concrete structure (other than a building), any road, airfield, sea defence works or river works and any other civil or construction engineering works of a similar nature. Also now included is the construction , structural alteration or repair (including re-painting) or the demolition of any pipe-line other than water pipe-lines, except where carried on upon a railway or tramway.

OBLIGATIONS

Regulation 3 in each of the General Provisions, Working Places, and Lifting Operations Regulations provides details of the general duties of every contractor, and every employer of workmen, who undertakes work to which the Regulations apply. They also contain the general duties owed by employees working

Introduction to the Construction Regulations

in the industry, which are to comply with those Regulations which require something to be done or not done, and to report any defect in plant or equipment without unreasonable delay.

Contractors and employers of workmen are required to comply with those Regulations which affect him or any of his employees, and if they install, erect, work or use any plant or equipment or erect or alter scaffolding, this must be done in compliance with the relevant Regulations. Under the Construction Regulations generally, (but not at common law) there is no obligation owed to employees of subcontractors. Obligations to workmen do not apply if their presence on the site is not authorised or permitted, or in the course of carrying out work for their employer.

Work is classed as being done by the main contractor if it is being done by a subcontractor and the main contractor retains control over it (Donaghey v Boulton and Paul Ltd [1968] AC 1, [1967] 2 All ER 1014).

OBJECTIVES OF THE REGULATIONS

The Regulations were designed to require protection against the specific dangers arising from construction work. The types of dangers that arise are many and varied, but accident records show that most construction workers are killed or injured as a result of falls from heights or during high risk activities such as demolition, roof work and excavation. It is in these areas that the

Regulations set standards creating a duty of compliance with that standard. Little information is available on the more general occupational health risks of construction work. In particular areas much research has been carried out and used to formulate control measures contained in legislation affecting industry as a whole in relation to specific hazards, such as asbestos.

The Chief Inspector of Factories may grant exemption from any or all of the requirements to any plant or equipment, or to any work.

HOW ARE THE OBJECTIVES MET?

The Regulations set standards of compliance in the five areas of general requirements, lifting operations, working places, health and welfare and head protection. Each of these will be summarised in turn. A major practical difficulty is that since the introduction of the first four Regulations in 1961 and 1966, research and experience has shown that an underlying cause of accidents in the industry has been lack of organisational control and clear lines of responsibility touching all parties from the client to the employee. These matters are not addressed by the first four sets of Regulations, apart from setting out the obligations of contractors and employers at the start of each. Proposals covering future requirements for the management aspects of construction work, to comply with the EC Temporary and Mobile Construction Sites Directive, are at a consultative stage at the time of writing, and will become the Construction Design and Management Regulations 1993.

The Construction (General Provisions) Regulations 1961

INTRODUCTION

The General Provisions Regulations contain the bulk of requirements for the general conduct of construction work. Work places and lifting operations are dealt with in other Regulations, as is site welfare. They introduced some important concepts, notably the requirement for larger firms in the industry to appoint one or more safety supervisors, and became effective on 1st March 1962.

OBJECTIVE OF THE REGULATIONS

The general objective of the Regulations is to promote the safe conduct of construction work.

HOW IS THE OBJECTIVE MET?

The main "software" requirement of these Regulations is designed to meet the objective by ensuring that employers and contractors working on construction activities appoint an experienced and adequately qualified person as safety supervisor to advise them on the requirements, and to supervise the observance of the Regulations and the safe conduct of the work generally. This must be done if the employer or contractor employs more than twenty people on construction work, not necessarily on the same site or all at work at once (**Regulation 5**).

Anyone given these duties must not have such a workload as to prevent the carrying out of the duties with reasonable efficiency. The name of anyone so appointed must be entered on the displayed copy of the Regulations. The Regulations do not prevent the same person being responsible for several sites, or a number of employers appointing the same person(s) (**Regulation 6**).

EXCAVATION WORK

Safety in excavation work is dealt with in **Regulations 8 – 14, 21 and 37**. Where necessary, an adequate supply of support material of suitable quality is to be provided and used as early as practicable to prevent the fall of material from sides of excavations, tunnel roofs and the like. An inspection requirement is provided by **Regulation 9** where people are working in excavations, shafts, earthworks or tunnels (which will all be classed as "excavations" hereafter). A competent person must inspect at least once each day, and if the depth exceeds 2m, the inspection is required at the beginning of every shift. Additionally, nobody may work in an excavation unless it has been thoroughly examined by a competent person within the preceding seven days, or if support work has been substantially damaged by unexpected falls of materials or by use of explosives which may have affected it.

Details of the inspection must normally be entered into the Register, Form F91 Part Section B, and be signed by the person making the inspection. This Regulation does not apply if persons are at work in the excavation installing support, as long as appropriate precautions are taken, and if the nature of the sides of the excavation is such that no-one working in it is liable to be buried, trapped or struck by materials falling more than 1.2m.

Regulation 10 requires that the installation of support work for any excavation is to be supervised by a competent person and carried out by competent, experienced workers. Materials used must be inspected by a competent person and any defective material found is not to be used. Support work must be soundly constructed and free from defect, adequate for the intended purpose and properly maintained. All struts and braces have to be secured so that they cannot be displaced accidentally.

Means of escape from excavations has to be provided if there is any danger of flooding (**Regulation 11**), but safe access into and out of excavations should always be available as required by the Construction (Working Places) Regulations.

Where an excavation could affect the stability of a structure, appropriate steps must be taken to ensure the safety of all those employed (**Regulation 12**). Every accessible part of an excavation or other opening in the ground near where people are working, where falls of 2m or more could occur, must be provided with a suitable barrier or a secure cover (**Regulation 13**) but this does not apply while the barrier or cover must be removed to carry out work or during the period between making a new opening and the first reasonably practicable opportunity to erect a barrier or cover. It should be noted that Sections 2 and 3 of the Health and Safety at Work etc. Act 1974 may be applied to extend this requirement beyond the situation where people are working nearby to include foreseeable approach by anyone, including the public.

Regulation 14 is designed to safeguard the edges of excavations by requiring that material is not placed there. No equipment or material should be placed or moved near an excavation if this might give rise to danger by causing the sides to collapse. **Regulation 21** requires all necessary action to be taken to ensure that any excavation or other confined space and the approaches to it is adequately ventilated so that the atmosphere is fit to breathe and any dangerous impurities in the air are rendered harmless. If it is suspected that the atmosphere there is toxic or asphyxiating, it must be tested by or under the supervision of a competent person. Work can only recommence when the atmosphere has been tested and shown to be safe.

If material is to be tipped into an excavation from a vehicle, measures must be taken under the requirement of **Regulation 37** to prevent the vehicle from running over the edge of the excavation.

COFFERDAMS AND CAISSONS

These are covered by **Regulations 15 – 18**. Their requirements are essentially the same as for excavations – proper construction of suitable and sound material free of defect. Each structure must be of adequate strength and properly maintained (**Regulation 15**). In every cofferdam and caisson there must be provided an adequate means of reaching places of safety in the event of flooding (**Regulation 16**). Construction, alteration and dismantling must be done under the immediate supervision of a competent person, and only carried out by competent workmen so far as possible, who possess adequate experience of such work. Materials used for construct on must be inspected for defects by a competent person on each occasion before being used. Defective or unsuitable material must not be used (**Regulation 17**). Similar inspection requirements to those for excavations are imposed by **Regulation 18**. The results of such examinations and inspections must be recorded in Form F91 Part 1 Section B.

EXPLOSIVES

Regulation 19 requires that explosives must not be handled or used except by or under the immediate control of a competent person. When explosives are fired, all those employed must not be exposed to risks of injury from the explosion or flying material, and steps must be taken to ensure this.

TRANSPORT

Proper arrangements are to be made for the conveyance of workers when they must work by water. Vessels used for transport must be suitably constructed, properly maintained, in the charge of a competent person and must neither be overloaded nor overcrowded (**Regulation 23**).

Any mechanically-propelled vehicles or mechanically-drawn trailers used to carry workers, goods or materials must be in good working order, mechanically sound, used in a proper manner and not loaded so that the vehicle cannot be operated safely (**Regulation 34**). Exceptions are made for vehicles which have broken down or have been damaged, as long as specified requirements are met. The Regulation does not apply to locomotives, trucks or wagons. **Regulation 35** prohibits riding on any mechanically-propelled or -drawn vehicle in an insecure position. They are only to use the place provided for that purpose. **Regulation 36** prohibits anyone from remaining on a vehicle if it is dangerous to do so, while it is being loaded with loose materials using a mechanical loader.

Rails and rail tracks are covered by **Regulation 25**; **Regulation 26** deals with the maintenance of locomotives, trucks and wagons. **Regulation 27** requires provision of adequate clearance so that persons cannot be crushed or trapped by any of these whilst passing by. Gantries and elevated structures carrying rails with traffic must have an adequate footway if people are required to walk along it (**Regulation 28**). If the footway is on the outside of the rail track and any person is likely to fall more than 2m from it, suitable guard rails (not less than 1m high) must be provided. Track vehicles require brakes fitting, except trucks or wagons used in circumstances which render a brake unnecessary. A sufficient number of suitable sprags or scotches must be provided for use in the movement of trucks or wagons. Provision of equipment for replacing derailed equipment, fitting of audible warning devices to locomotives, and the giving of adequate warning by those in charge of movements are all covered in the Regulations.

Only trained and competent people over the age of eighteen are allowed to drive mechanically-propelled vehicles and some other pieces of heavy equipment, unless direct supervision of a qualified person whilst under training (**Regulation 32**).

DEMOLITION

The Regulations apply to the demolition of the whole or a substantial part of a structure or building, and where this is done a competent supervisor must be appointed who is experienced in the work (**Regulation 39**). Further, if more than one contractor is involved, they should either appoint their own supervisors or a joint one. However, if several are to oversee the work the contractors must ensure that a plan of operation is

Module D Section 21

The Construction (General Provisions) Regulations 1961

jointly agreed before work begins. If the demolition work is to be done by individual contractors, then each must consult with the others before any operation starts, and no operation can be undertaken until a method and timetable has been jointly agreed.

The required precautions to be taken are set out at some length in **Regulations 40 and 41**. All practicable precautions must be taken before work starts and while it is in progress to prevent those involved being endangered by fire or explosion resulting from a leak or accumulation of gas or vapour, and by flooding.

The main provisions of Regulation 41 are:

- No part of a building or structure can be allowed to become dangerously overloaded with debris
- The following work must be done under the close supervision of an competent supervisor with adequate experience of the work, or by experienced workers with the direction of a competent supervisor:
  a) demolition of all or part of a building or structure (except where there is no foreseeable risk of collapse
  b) demolition of all or part of a building or structure where there **is** a special risk of collapse of anything as a result of the demolition
  c) the cutting of any reinforced concrete steelwork or ironwork forming part of the building or structure being demolished, and before the cutting precautions are to be taken to prevent twisting, springing or collapse
- When part of a frame is removed from a framed or partly-framed building or structure, all practicable precautions are to be taken to prevent collapse
- Both before demolition begins and while it is in progress, all practicable precautions are required to prevent employees being endangered by the accidental collapse of any part of the building, structure or anything adjoining it.

SAFE WORKING ENVIRONMENT

Further General Provisions Regulations are designed to provide a safe working environment, and deal with a wide variety of hazards in construction work which are not dealt with by the other Construction Regulations.

Appropriate steps are to be taken if there is a risk of drowning associated with the work (**Regulation 24**). Suitable rescue equipment must be provided and maintained ready for immediate use and in an efficient state. Where there is a special risk, all edges from which a person may fall with a subsequent risk of drowning

must be securely fenced. Fencing may be removed to allow the passage of materials or persons only, and then must be replaced promptly.

Measures are required to prevent any steam, smoke or vapour generated on site from obscuring the area where people are working (**Regulation 45**).

Precautions must be taken to prevent anyone being struck by a falling object while working at any place on the site which is a habitual or contemplated place of employment (**Regulation 46**). Materials and waste must not be thrown down from scaffolds where they are liable to cause injury, but are to be lowered where practicable. Where it is not, adequate precautions are required to prevent employees being struck by falling objects or flying debris.

Lighting of an adequate and suitable kind is required for all working places, access routes to them, places where lifting appliances operate and all dangerous openings (**Regulation 47**).

Timber (or any other material) which contains protruding nails cannot be used in any work where it might cause harm. **Regulation 48** continues to require it to be removed from any place where it might cause harm to employees. Safe stacking of materials and removal of obstructing material from access routes is also covered.

Painted or cement-washed ironwork or steelwork can be moved, handled, walked on or walked on until the paint or wash on it is dry. This, **Regulation 51**, does not apply to the movement or handling of these articles in order to apply paint or cement wash.

For pre-1993 equipment only, every dangerous part of machinery, mechanically-powered or otherwise, must be securely fenced, by **Regulation 42**. This is not necessary if the machine is constructed or positioned so that it cannot expose anyone to danger. For post-1992 equipment, the Provision and Use of Work Equipment Regulations 1992 apply.

Temporary structures, except scaffolds and others to which Regulation 11 of the Construction (Lifting Operations) Regulations applies, must be of good construction, adequate strength and stability, of sound material and properly maintained (**Regulation 49**). Structures covered by this requirement include bridges, storage racks and the support for shuttering used in reinforced concrete work.

Building structures which become temporarily unstable during any phase of construction must be adequately supported by appropriate means to prevent collapse (**Regulation 50**). This would apply to structures or buildings made unstable by some work operation, but does not apply to demolition work.

Pile driving helmets and crowns are required to be of good construction and of sound, suitable material of adequate strength and free from defect (**Regulation 53**).

RECORD KEEPING

Reports relating to the examinations of excavations, cofferdams and caissons are to be kept on site until the work is complete. After completion of the work they are to be kept at the office of the contractor ordering the examinations. Where operations are expected to last for less than six weeks the reports can be kept at the contractor's office (**Regulation 56**). All reports must be made available for inspection by a Factory Inspector, and copies must be sent to the Inspector on request.

Revision

The Construction (General Provisions) Regulations require:

- Supervision of safe conduct of work
- Compliance with listed requirements on:
    Excavations
    Cofferdams and caissons
    Explosives
    Work on or near water
    Transport
    Demolition
- Compliance with listed general precautions.

The Construction (Lifting Operations) Regulations 1961

INTRODUCTION

The Provision and Use of Work Equipment Regulations 1992 are relevant to lifting equipment first provided for use after January 1993, in particular Regulations 5, 6, 9, 14, 20, 23 and 24. These refer to suitability, maintenance, training, operating controls, stability, markings and warnings respectively. In addition, the Lifting Operations Regulations contain the detailed requirements designed to prevent accidents during lifting operations on site. They also apply in full to older equipment.

The flexibility and ease of use of modern lifting appliances can lead to misuse and complacency which in turn result in accidents and dangerous occurrences. Section 2 of the Safety Technology Module covers some aspects of mechanical handling. These Regulations came into effect on 1st March 1962, and apply to all building operations and works of engineering construction. See the Introduction to the Construction Regulations for definitions of these terms, and duties of employers and contractors.

This summary is not intended to be an authoritative interpretation, and the reader is recommended to read this Section in conjunction with a copy of the Regulations.

Some important **definitions** of terms are given in the Regulations. These include:

'Lifting appliance', which is defined as a crab, winch, pulley block or gin wheel used for raising or lowering, and a hoist, crane, sheerlegs excavator, dragline, piling frame, aerial cableway, aerial ropeway or overhead runway.

'Lifting gear' means a chain sling, rope sling or similar gear, and a ring, link, hook, plate clamp, shackle, swivel or eyebolt.

OBJECTIVES OF THE REGULATIONS

The general objectives of the Regulations are to control and eliminate the main causes of accidents associated with lifting operations on site. These include lack of planning, training and maintenance – in particular selecting the wrong type of equipment, siting of lifting equipment, use of wrong lifting gear, failure of personnel to carry out the correct procedures, lack of proper maintenance, and absence of properly-trained personnel.

HOW ARE THE OBJECTIVES MET?

Regulation 5 permits the Chief Inspector to issue written certificates to exempt plant or equipment from the requirements of the Regulations. Further exemptions are made under **Regulation 6** to lifting machinery in factory premises and in docks. Under **Regulation 7** some further exemptions are granted to the chains, ropes and lifting gear which arrive on site attached to a load intended for use in work on construction sites.

SAFETY OF LIFTING APPLIANCES

Regulation 10 requires every lifting appliance and all associated equipment to be well-constructed, free from defect, suitable for the purpose, properly maintained, and inspected once a week by a competent person who is to sign a report (Form F91 Part 1 Section C). Every lifting appliance must be adequately and securely supported (**Regulation 11**). Anchoring and fixing arrangements must be adequate and secure.

Regulation 12 requires safe clearances to be left around lifting appliances which have a travelling or slewing motion to ensure safe access of people to avoid crushing. Platforms from which crane drivers and signallers have to work must be of suitable area, closeboarded, have a safe means of access to them and be provided with edge protection if a fall of more than 2m is possible (**Regulation 13**). Cabs for drivers of all kinds of lifting appliances must protect them from the weather (**Regulation 14**).

Only competent persons over the age of 18 are permitted to operate lifting appliances, unless under direct supervision for training, and similarly those under 18 are not allowed to give signals to drivers of lifting appliances (**Regulation 26**). All signals given are to be distinctive and easily seen or heard (**Regulation 27**).

A safe means of access must be provided so far as is reasonably practicable where a person is required to examine, repair or lubricate a lifting appliance and can fall more than 2m (**Regulation 17**).

ASSOCIATED EQUIPMENT

Winding drums and pulleys must be adequate for the purpose, and diameters must be suitable for ropes or chains to be used. These must be secured to winding drums, and at every operating position there must be at least two turns remaining on the drum (**Regulation 15**). **Regulation 16** provides that efficient brakes and safety devices are used to prevent a load falling and to control its descent. Controls must be marked, and provided with some locking arrangement if accidental displacement could result in danger.

SUPPORTS FOR PULLEY BLOCKS AND GIN WHEELS

Regulation 18 requires that pulley blocks and gin wheels must not be used unless they are effectively secured to the poles or beams they are suspended from. Poles and beams must be of adequate strength and appropriately secured.

CHAINS, ROPES AND LIFTING GEAR

Chains, ropes and lifting gear is to be of sound construction, free from defects and suitable for the job. They must be tested and examined (except fibre ropes or slings) with appropriate certificates issued. They must be marked with means of identification and their Safe Working Load (SWL). Over-loading past the SWL is forbidden. Alternatives for the marking or ropes and rope slings are allowed (putting the SWL in a report or table). No wire rope is to be used if the number of broken wires visible in a 10-diameter length of it exceeds 5% of the total number of wires in the rope (**Regulation 34**).

A load must not be raised, lowered or suspended by a chain or rope which has a knot in any part under tension (**Regulation 39**). Further, chains which have been shortened or joined by means of bolts inserted through the links must not be used. Chains, ropes and other lifting gear must be thoroughly examined once every six months. Those not in "regular use" can be examined as necessary. Reports must be completed and signed (Form F91 Part 2 Section 1) (**Regulation 40**).

Chains and ropes which have been lengthened, altered or repaired by welding must be thoroughly examined and tested for SWL. A report is to be completed and signed (Form F97) (**Regulation 35**). Chains attached to the buckets of draglines or excavators are exempt from this requirement. Annealing (or other heat treatment) of chains and lifting gear, other than rope slings or lifting gear specified in the Second Schedule to the Regulations or those exempted by certificate, must be carried out every 14 months, with the work supervised by a competent person. This must be done every 6 months if the gear is made from 13mm bar or smaller (**Regulation 41**). Heat treatment does not apply to lifting gear not in regular use, or which is used solely on hand-powered lifting appliances. Annealing etc. is required only as necessary in these circumstances. A written report is required once heat treatment has been carried out, signed by a competent person.

LIFTING OPERATIONS

Every hook must be fitted with a safety catch, or be shaped so as to prevent the load being displaced (**Regulation 36**). Slings which are used must be securely attached to the appliance to prevent damage. Double or multiple slings must not be used if the upper ends of the sling legs are not connected by means of a shackle, ring or other device of adequate strength, or if the angle between the sling legs is such that the SWL of any of the sling legs is exceeded (**Regulation 37**).

RAISING AND LOWERING OF LOADS

The SWL and identification must be clearly marked on every crane, crab or winch; every pulley block, gin wheel, sheerlegs, derrick pole, derrick mast or aerial cableway used for lowering or raising loads of one tone or more. Every crane of variable operating radius must have the SWL plainly marked at the various radii. For cranes with derricking jibs the maximum radius at which the jib may be operated is also to be shown (**Regulation 29**).

The SWL of the appliances noted above must not be exceeded except when being tested (**Regulation 31**).

When lifting loads which are close to the SWL of the appliance, the lift must be halted after the load has been raised a short distance before proceeding with the operation (cranes, crabs and winches only). Tandem lifting is also controlled by this Regulation, **32**.

Precautions are to be taken to prevent the edges of loads coming into contact with any slings etc. if this could cause danger (**Regulation 38**). Specifications to ensure the secureness of loads being lifted are made in **Regulation 49**, and **Regulation 47** provides controls over the carriage of persons by means of lifting appliances.

HOISTS

Safety at hoistways is ensured by the requirements to provide gates, enclosures, arrestor gear and over-run devices (**Regulation 42**). Hoists must only be operated from one position where practicable, and if the operator's view is restricted a signalling system must be used to allow the operator to stop the platform at the appropriate level. Hoist winches must be fitted with 'dead man' brakes (**Regulation 44**) and the SWL is to be marked on the hoist (**Regulation 45**). Further requirements are placed on passenger-carrying hoists, which must be operated from within the cage. Test and examination procedures are described in **Regulation 46**. Certain exemptions for permanently installed hoists

The Construction (Lifting Operations) Regulations 1961

from the requirements of the Regulations are granted by **Regulation 8**.

CRANES

Stability of lifting appliances is covered by **Regulation 19**, and **Regulation 20** sets out requirements for rail-mounted cranes. Bogies, trolleys or wheeled carriages on which cranes are mounted must be suitable for the purpose (**Regulation 21**). Restrictions are placed on the use of cranes by **Regulation 23**, which forbids their use for any purpose other than raising or lowering a load unless properly supervised, the stability is not put at risk and no part of the structure or mechanism has undue stress imposed upon it. Timber structural members are not permitted (**Regulation 24**).

Cranes are to be erected only under supervision of competent persons (**Regulation 25**), and testing and examination requirements are set out in **Regulation 28**. **Regulation 30** makes provision for the indication of the SWL on jib cranes by means of automatic safe load indicators (ASLI). Scotch and guy derrick cranes have their own peculiarities, controlled by **Regulation 33**.

REVISION

The Construction (Lifting Operations) Regulations cover:

- Lifting appliances of all types used in construction
- Chains, ropes and lifting gear
- Special requirements for hoists
- Carriage of workers
- Secureness of loads
- Record-keeping requirements.

The Construction (Working Places) Regulations 1966

INTRODUCTION

The Provision and Use of Work Equipment Regulations 1992 are relevant to some aspects of the Working Places Regulations, where equipment is erected or first provided for use after January 1993, in particular Regulations 5, 6, 9 and 20. These refer to suitability of work equipment, maintenance, training and stability. The Workplace (Health, Safety and Welfare) Regulations do not apply to construction sites.

The Working Places Regulations contain the detailed requirements for safety of working places in the construction industry, particularly scaffolding. Section 4 of the Safety Technology Module covers access equipment. The Regulations came into effect on 1st August 1966, and apply to all building operations and works of engineering construction. See the Introduction to the Construction Regulations for definitions of these terms, and duties of employers and contractors.

This summary is not intended to be an authoritative interpretation, and the reader is recommended to read this Section in conjunction with a copy of the Regulations.

SOME IMPORTANT DEFINITIONS

Ladder – the term includes a timber or metal ladder or stepladder, but not a pair of steps

Scaffold – any temporary working platform and its supports; a single plank resting on roof framework, or a specially designed platform on shuttering would be included

Slung scaffold – a working platform hung on ropes, tubes, poles etc. at a fixed height

Suspended scaffold – a 'cradle' type platform raised or lowered on pulley blocks or winch gear, but not a boatswain's (bosun's) chair

Trestle scaffold – a platform rested on A frames or folding supports such as painters' trestles, pairs of steps

Working platform – any platform from which a worker works.

OBJECTIVE OF THE REGULATIONS

The general objective of the Regulations is to promote the safe use of access equipment, including use of proper foundations, sound materials and avoiding structural failures. Poor design of working areas at heights and falls through lack of protection can lead to injury and death, and are controlled by these Regulations.

HOW IS THE OBJECTIVE MET?

Regulation 5 permits the Chief Inspector to grant exemptions from the requirements to plant, equipment or particular work – some are granted to external scaffolds for steeplejack work, for example.

SAFE PLACE OF WORK AND PROVISION OF SCAFFOLDS

A suitable, adequate and safe means of getting to and from a place of work must be provided, so far as is reasonably practicable, under **Regulation 6**. The means of access is to be properly maintained, and be in position before the work commences. Equally, every place where someone is required to work must be made and kept safe, so far as is reasonably practicable. If work cannot be carried out safely from the ground or from a building or other permanent structure, a scaffold, or ladder if appropriate, must be provided which is suitable and sufficient, and which is properly maintained (**Regulation 7**). The scaffold must be properly placed, erected and secured for use.

The erection, dismantling, addition to or alteration of scaffolds must be under the immediate supervision of a competent person (**Regulation 8**). Workers involved in this work are to be competent and experienced where possible. Scaffolding materials must be inspected by a competent person on each occasion before use.

SCAFFOLD ERECTION AND USE

Scaffolds and their components must be well constructed from suitable and sound material, and be of adequate strength. Sufficient materials must be available to construct the scaffold. Timber used must be of good quality and condition; all bark must be removed, and it must not be painted or treated so as to hide defects. Metal components for scaffolds must be of good quality and condition, free from corrosion or any other defects which would impair their strength (**Regulation 9**).

Defective materials or components must not be used to construct scaffolds, and defective ropes or bonds must not be used. All sound scaffold material must be properly stored when not in use, and must be kept apart from any defective material (**Regulation 10**). Every scaffold must be properly maintained, and accidental

The Construction (Working Places) Regulations 1966

displacement must be prevented so far as is reasonably practicable by ensuring that they are securely placed or fixed in position (**Regulation 11**).

Scaffolds or parts of scaffolds must not remain partly erected or dismantled unless they comply with these Regulations or a notice is prominently displayed warning that it must not be used. Alternatively, access may be blocked so far as is reasonably practicable (**Regulation 12**).

Scaffold standards (uprights) must where practicable be either vertical or slightly inclined towards the building or structure, and placed sufficiently close together to ensure that the scaffold is stable. **Regulation 13** also requires base plates or other means to be provided to prevent slipping or sinking; ledgers must be as close to the horizontal as possible and secured to the standards by an efficient means; and putlogs or other supports must be securely fastened to uprights or their equivalent.

Putlogs are short tubes which can support scaffold boards, with one flattened end or end fitting called a spade. Where they are supported by a wall, a large enough area of the flat part must be inserted into the wall to provide support. Where scaffolds are independent, (not depending directly upon a building or structure for support) the short tubes are known as transoms. The distance between adjacent putlogs or transoms must be adequate to support the platform and the loads they carry. Where a platform is a single thickness of board these distances in general must not exceed 1.0m for boards 32mm thick, 1.5m for boards 38mm thick and 2.6m for boards 50mm thick.

Regulation 14 requires that any ladders used as scaffold uprights must be strong enough, placed so the two sides or stiles are evenly supported or suspended, and secured so that they cannot slip. Ladder scaffolds can only be used for light work such as painting, and where the material is such that the scaffold can be used safely.

Stability of scaffolds is covered in **Regulation 15**, for independent, putlog or mobile units, dealing with the provision of braces, ties, the structural integrity of the building supporting the scaffold, and prohibition of the use of loose bricks or blocks, drain pipes etc. as support. A number of requirements are made concerning the construction, stability and firm base requirements for mobile towers.

SUSPENDED SCAFFOLDS

Regulation 16 requires that slung scaffolds comply with all the Regulations and that suspension arrange-

ments (chains, wire ropes, lifting gear etc.) comply with requirements which ensure the integrity of the system. **Regulation 17** gives details about the use of cantilever and jib scaffolds.

Figure and bracket scaffolds cannot be used if there is a danger that their fixings can be pulled out of the brickwork. Parts of buildings used in the support of scaffolds, ladders, folding stepladders or crawling boards must be of adequate strength and stability (**Regulation 18**).

The safe use of suspended scaffolds which are not power-operated is dealt with by **Regulation 19**, and **Regulation 20** notes further requirements for the use of boatswain's (bosun's) chairs, cages, skips etc. which are also not power-operated.

TRESTLE SCAFFOLDS

Trestle scaffolds are to be well-constructed from suitable equipment which is free from defect. They are not to be used if anyone can fall more than 4.5m from them, or if more than one tier rests on folding supports (**Regulation 21**). Trestle scaffolds must not be erected on a scaffold platform unless there is clear space for the passage of people and materials and the trestle supports are fixed to the platform and braced to prevent accidental movement.

INSPECTIONS

Regulation 22 deals with inspection requirements. No scaffold, boatswain's chair or similar equipment is to be used unless it has been inspected by a competent person within the preceding seven days, or since it was exposed to weather conditions which may have affected it, and the results have been recorded (Form F91 Part 1 Section A). The reports are to be kept on site unless the work is to last less than six weeks, and must be made available to an Inspector on demand (**Regulation 39**). Where a part of a scaffold is to be used by another employer's workers, the second employer must satisfy himself that the scaffold is stable and sound, and provided with all safeguards, before his workers use it (**Regulation 23**).

ACCESS

Regulation 24 requires a working platform to be free from tripping hazards and slipping hazards by close boarding and prevention of excessive slopes. **Regulation 25** specifies the boards and planks to be used in gangways, working platforms and runs, and the ways in which they should be placed and supported to ensure safety. **Regulation 26** specifies dimensions to ensure an adequate width of working platforms to ensure the safe access of persons and materials. Similar width requirements for gangways and runs are contained in **Regulation 27**, while **Regulations 28 and 29** provide for fitting of toeboards and guardrails to scaffolds and gangways and runs, from which persons or materials can fall a distance of more than 2m. Platforms, gangways, runs and stairs must give a safe foothold, be clear of rubbish and not slippery (**Regulation 30**).

LADDERS

Regulation 31 deals with ladder construction. Every ladder or folding stepladder is to be well-made from suitable material, of adequate strength and properly maintained. A ladder must not be used if a rung is missing or defective. Rungs must be securely fastened to the stiles (except crawling ladders). **Regulation 32** places requirements on use of ladders and stepladders. They must be tied or footed on a firm level base, and be equally supported on both stiles. Ladders must extend at least 1.05m above the landing place or have adequate handholds either side.

Guardrails and toeboards must be provided to all edges from which people can fall in excess of 2m, or which are over water, liquid or dangerous material. All holes in floors must be protected or covered over. Open joisting should also be covered if there is a risk of a fall of 2m (**Regulation 33**). Guardrails etc. can be removed for the access of workers or materials, but must be replaced as soon as possible afterwards (**Regulation 34**).

ROOFWORK

A number of precautions are required by **Regulations 35** when working on sloping roofs, or on or near fragile materials (**36**). **Regulation 38** provides for the use of safety nets, belts or harnesses. These state in summary that only suitable workers are to be used for work on sloping roofs, using crawling boards or ladders if the material is fragile. Safety harnesses can be used as alternatives only when other means of fall protection are not practicable. **Regulation 37** forbids the overloading of scaffolds – loads must be evenly distributed, with loading operations done with care to avoid the imposition of shock loads. Materials may only be kept on scaffolds if they are required for use in a reasonable time.

REVISION

The Construction (Working Places) Regulations cover:

- Scaffolds and safe access to working places
- Ladders
- Different types of scaffolds

- Inspection and maintenance of scaffolds
- Details for working platforms, gangways and runs
- Prevention of falls at platforms and openings
- Roofwork.

The Construction (Health and Welfare) Regulations 1966

INTRODUCTION

In force well before the Health and Safety at Work etc. Act 1974, this set of Regulations is typical of the old approach to health and safety issues, especially in that it provides many numerical standards to meet. After 1975, legislation made on a subject rather than an industry basis revoked the first-aid provisions (the majority of the Regulations). They became effective on 1st May 1966.

OBJECTIVES OF THE REGULATIONS

The objectives of the Regulations are to ensure the provision of adequate facilities on site for the health and welfare of employees.

HOW ARE THE OBJECTIVES MET?

By **Regulation 4**, every contractor or employer of workmen has to comply with the Regulations which affect his employees. Exemptions to this can be made by the Chief Inspector of Factories. Each contractor or employer can arrange to share the facilities required by Regulations 11, 12 and 13 provided by another contractor on site, or any other person. The facilities must be adequate for the combined numbers, and convenient to the work area. The contractor providing the shared facilities must enter the details of the shared facilities in Form F2202, and give the contractor to whom the facilities are provided a copy (Form F2202A) containing the details of the facilities.

This should be done as soon as the arrangements are made. The certificates and copies should be kept on site or at the contractor's office and made available for inspection by Inspectors or any employee affected by them at all reasonable times.

Regulation 11 requires the provision of shelter against bad weather, which also contains storage for personal clothes, and either a) an adequate source of heat for warmth and drying if more than five people are employed, or b) arrangements for providing warmth and drying facilities, so far as is reasonably practicable, if five or fewer are employed.

Also required on all sites is convenient storage for protective working clothes kept on site, and such facilities for drying them as is reasonably practicable. Suitable accommodation for eating food, provided with sufficient tables and seats and with facilities for boiling water is required regardless of numbers employed, but where more than ten are employed then facilities for heating food are required if hot food is not available on site.

The adequacy of this accommodation can be judged according to the numbers who appear to be likely to use it at any particular time. The accommodation is to be kept clean, and is not to be used for the storage of plant or materials. In the accommodation or elsewhere on site, an adequate supply of drinking water must be installed, clearly marked as such if this is not obvious.

Regulation 12 deals with the provision of washing facilities, and requires these if at least one employee is on site for four consecutive hours. If a contractor has more than twenty employees on site and believes the work will not be completed within six weeks, he must provide water containers (basins, buckets or troughs with smooth impervious internal surfaces), an adequate supply of soap and towels, and sufficient hot and cold or warm water.

Where a contractor has more than 100 employees on site and the work will not be completed within twelve months, he must provide four wash-basins. One further basin is required for each unit of 35 in excess of 100, and parts of a unit are rounded up to the nearest whole unit. An adequate supply of hot and cold or warm water, soap and towels (or other means of hand-drying) must also be supplied. If lead or other poisonous substances are used on site, extra provision must be made including the provision of nail brushes and water containers on the scale of one for every five people using the material. Washing facilities must be conveniently accessible from the accommodation provided for taking meals, and must be kept clean.

The provision on every site of at least one suitable toilet, not a urinal, is required for every 25 employees on site. If more than 100 are present on site, and sufficient urinals are also provided, then one toilet for every 25 employees is adequate up to the first 100, with one extra for every unit of 35 employees thereafter. Again, parts of a unit count as a whole unit. Water closets (flushing toilets) are regarded as suitable, but chemical toilets should not be used unless there is no practicable alternative and adequate arrangements for emptying have been made.

Regulation 14 requires every toilet to be well-ventilated, and not open directly into a workroom or messroom. Every toilet other than a urinal must be covered, and fitted with a proper door and fastening. Urinals must be screened from view both on and off the site. Toilet facilities are to be kept clean and well lit, as well as conveniently accessible to the site. Separate toilets are required for male and female employees.

The Construction (Health and Welfare) Regulations 1966

Regulation 15, on issue of protective clothing, has been revoked in favour of the Personal Protective Equipment at Work Regulations 1992, all of which apply to construction work. Facilities provided and their means of access must be safe for persons using them, so far as is reasonably practicable (**Regulation 16**).

REVISION

The Construction (Health and Welfare) Regulations 1966 require the provision of:

- Shelters and accommodation for clothing and taking meals
- Washing facilities
- Sanitary conveniences
- Safe access to facilities.

The Construction (Head Protection) Regulations 1989

INTRODUCTION

Accident statistics for the construction industry indicate that a small but significant number of injuries are caused by failure to prevent objects falling from heights, and by operations with lifting appliances. Injuries involving the head are obviously more serious. A number of activities are also regarded as high-risk, notably demolition and excavation. The industry has had a poor experience of voluntary use of safety helmets, so that legislative control finally became necessary. This was also welcomed by the industry because of its standardising effect which would increase the acceptability of controls by the workforce. Early indications are that up to 250 construction workers may have been saved from head injury in the first year of operation of the Regulations, according to a Health and Safety Commission statement in December 1991.

OBJECTIVES OF THE REGULATIONS

The objectives of the Regulations are to ensure the provision, maintenance and use of adequate head protection on construction work.

HOW ARE THE OBJECTIVES MET?

The basic principle of the Regulations is to require everyone except turban-wearing Sikhs working in the industry to wear suitable head protection whenever there is a risk of injury to the head from falling objects or hitting the head against something. The head protection is not required to protect against head injury caused by a fall of a person. The Regulations came into force on 30th March 1990.

These Regulations apply to the same activities as the other Construction Regulations (**Regulation 2**), but not to diving operations.

Regulation 1 defines what is meant by "suitable" head protection – it is that which:

- is designed to provide protection so far as is reasonably practicable against the risk of head injury – this means conformance to current standards, presently BS 5240. The risk must be foreseeable given the environment in which the work takes place.
- fits the wearer, after suitable adjustment
- is suitable for the activity in which the wearer may be engaged.

Sections 11 and 12 of the Employment Act 1989 have been used to exempt members of the Sikh religion who are wearing turbans from any requirement to wear head protection on a construction site. A Sikh not wearing a turban is required to comply with these Regulations in all respects. Sikhs choosing to wear turbans deny themselves use of adequate head protection, and provision is made to limit the employer's liability in the event of a claim, because of this.

Each employer must provide suitable head protection for each employee, and must ensure that it is properly maintained and replaced whenever necessary (**Regulation 3**). Self-employed persons must provide their own, and equally ensure its proper maintenance and replacement when necessary. Any head protection provided must comply with any legislation implementing provisions on design or manufacture with respect to health or safety in any relevant Community Directive applicable to head protection. Before choosing head protection, an employer or self-employed person must make an assessment to determine whether it is suitable (**Regulation 3(4)**).

The assessment must involve:

- the definition of the characteristics which head protection must have in order to be suitable
- comparison of the protection available with the above characteristics

The assessment must be reviewed if there is reason to suspect it is no longer valid, or if there has been a significant change in the work to which it relates (**Regulation 3(6)**). Every employer and self-employed person must ensure that appropriate accommodation is available for head protection when it is not being used (**Regulation 3(7)**).

Every employer is to ensure, so far as is reasonably practicable, that each of his employees wears suitable head protection, unless there is no foreseeable risk of injury to the head other than by falling (**Regulation 4**). Every person, employer, employee or self-employed person who has control over someone else is also responsible for seeing that head protection is worn where there is a foreseeable risk of injury (except from falling).

A person in charge of a site may make rules applying to the site, or give directions to ensure his employees comply with Regulation 4. This is a requirement of **Regulation 5**, which also requires these rules to be in writing, and to be brought to the attention of all those who may be affected by them. Once this is done, **Regulation 6** gives the rules the force of law, because it requires every employee to wear head protection when required to do so by rules or instructions. Similar duties are placed on the self-employed to comply with the

The Construction (Head Protection) Regulations 1989

instructions and rules of those in charge of the site. Full and proper use must be made of the head protection, so it must be worn properly, that is, as the manufacturer intended – not backwards, for example. Employees must also take all reasonable steps to return it to the accommodation provided for it after use (**Regulation 6(4)**).

Employees given head protection for their use must take

reasonable care of it, and report any loss or damage to their employer without delay (**Regulation 7**).

Exemptions may be granted to persons or activities by the Health and Safety Executive, but not if granting an exemption might have an adverse effect on the health and safety of those likely to be affected by the exemption (**Regulation 9**).

REVISION

The Construction (Head Protection) Regulations 1989 require:

- Provision and replacement by the employer of suitable head protection
- Suitability assessment
- The employer and those in charge to ensure it is worn where risks of head injury exist

- Compliance by employees with any rules the employer may make on wearing head protection in areas
- Accommodation to be provided
- The employee to make full and proper use of the head protection
- The employee to return it to its accommodation after use, and report loss or defects.

The Highly Flammable Liquids and Liquefied Petroleum Gases Regulations 1972

INTRODUCTION

These Regulations became fully effective on 21st June 1974, imposing requirements for the protection of persons employed in factories and other premises where the Factories Act 1961 applies and in which any highly flammable liquid or liquefied petroleum gas is present for the purpose of or in connection with any undertaking, trade or business. For liquefied petroleum gases the requirements are limited to the manner of storage and the marking of storage accommodation. Also applicable to circumstances where such substances are stored are additional Regulations such as the Dangerous Substances (Notification and Marking of Sites) Regulations 1990.

The Highly Flammable Liquids and Liquefied Petroleum Gases Regulations apply when liquids, with a flash point of less than 32 degrees Celsius and which support combustion when tested in the prescribed way, are present at premises subject to the Factories Act 1961. An exception to the storage requirements of these Regulations applies where a **petroleum licence** is in force – see below.

The Dangerous Substances (Conveyance by Road in Road Tankers and Tank Containers) Regulations 1981 and the Road Traffic (Carriage of Dangerous Substances in Packages etc.) Regulations 1986 apply to the loading, unloading and carriage of flammable liquids. Approved Codes of Practice give guidance on compliance with the operational provisions of both these sets of Regulations.

The Fire Precautions Act 1971 allows the presence of flammable liquids to be taken into account when considering general fire precautions.

The Petroleum (Consolidation) Act 1928 (as extended by the Petroleum (Mixtures) Order 1929 and other legislation) requires the keeping of petroleum spirit and/or mixtures to be authorised by licence and to be in accordance with any conditions attached to the licence.

The Petroleum Spirit (Motor Vehicles etc.) Regulations 1929 and the Petroleum Spirit (Plastic Containers) Regulations 1982 set conditions on the keeping of small amounts of petroleum spirit without a licence, which is intended for use in an internal combustion engine and wholly or partly for sale. Approved Code of Practice No.6 gives guidance on the testing and marking or labelling of plastic containers to contain up to 5 litres of petroleum spirit.

Also relevant to the storage, handling, transportation and use of highly flammable liquids and liquefied petroleum gases (LPG) are the Health and Safety at Work etc. Act 1974 Sections 2–4 and 6–8.

OBJECTIVES OF THE REGULATIONS

The Regulations were designed to obviate the risks associated with the storage of highly flammable liquids and liquefied petroleum gases.

HOW ARE THE OBJECTIVES MET?

The Regulations apply to all factories, and other premises to which the Factories Act 1961 is extended (**Regulation 3**). These include construction sites and warehouses. They impose duties on the occupier of the premises, or in some cases on the owner of the substance(s). They do not apply to premises subject to the Offices, Shops and Railway Premises Act 1963, or to research laboratories.

Highly flammable liquids (HFL) are defined by reference to flash point and combustibility; the method of test to determine whether a substance falls within the definition is set out in Schedules 1 and 2 of the Regulations. The flash point of 32 degrees Celsius was adopted after wide discussion at an inquiry into the draft Regulations in 1971. Both tests are necessary since certain mixtures of liquids can give a positive result in the flash point test (**Schedule 1**) and but not support combustion (**Schedule 2**). LPG is defined as commercial butane, commercial propane and any mixture of the two.

Regulation 5 is intended to ensure that all HFL, when not in actual use or being conveyed, is stored in a safe manner. The main purpose of the Regulation is to achieve the separation of storage from process areas, and to ensure so far as is reasonably practicable that the storage containers are sited in a safe position, preferably in the open air. There is also provision in the Regulation for the containment of any spillage from the containers.

In summary, HFL must be stored:

- In suitable fixed storage tanks in safe positions; or
- In suitable closed vessels kept in a safe position in the open air and, where necessary, protected against direct sunlight; or
- In suitable closed vessels kept in a storeroom which either is in a safe position or in a fire-resisting structure; or
- In the case of a workroom, where the combined total quantity of HFL stored is less than 50 litres, in suitable closed vessels kept in a suitably placed cupboard or bin which is itself a fire-resisting structure.

The Highly Flammable Liquids and Liquefied Petroleum Gases Regulations 1972

Regulation 6 requires the marking of storerooms, tanks, vessels etc. clearly and boldly with "Highly Flammable" or "Flash point below 32°C", or another appropriate indication of flammability. Where it is not practicable to mark in this way, the words "Highly Flammable Liquid" must be clearly and boldly displayed as near to the room, tank or vessel as possible. The purpose of marking is to ensure that people are aware of the flammable nature of the contents of the store or container.

The appropriate storage of LPG not in use is dealt with in **Regulation 7**, together with the appropriate marking of tanks, vessels and containers. In order to ensure that only the minimum amount is kept in any workplace, Regulation 7(3) specifies that LPG cylinders must be kept in a store until required for use, and any expended cylinder must be returned to the store facility as soon as is reasonably practicable. The effect of the requirement is that LPG cylinders which are not in use and connected to plant or equipment may remain in the workroom even though that plant or equipment is not in actual use. On the other hand, as far as is reasonably practicable the keeping of additional or unnecessary cylinders in the workroom is not permitted, whether the cylinders are full or have been used.

Precautions required against spills and leaks are covered by **Regulation 8**, whose principal object is the safe containment of HFLs and their vapours during such operations as transfer from storage to process and manipulation within the process area. Piped systems are preferred, but where not warranted suitable closed containers are permissible. Portable vessels when emptied must be removed without delay to a safe place.

Regulation 9 requires sources of ignition not to be present where dangerous concentrations of HFL vapours can be expected to be present, and the deposit of used cotton waste in closed containers if not removed without delay to a safe place.

Prevention of escape of vapours as far as is reasonably practicable is required by **Regulation 9**, which also covers exhaust ventilation for certain workrooms and its construction and maintenance, venting of fixed tanks and suitable placing of electric motors powering ventilation systems used in connection with HFLs. Provision of explosion relief for a fire-resisting structures can permit relaxation of the specification for the structure (**Regulation 11**), but the pressure relief must vent into a safe place.

For certain dock and warehouse premises and construction sites only, means of escape in case of fire is required to be provided which is both adequate and safe, in relation to every room in which HFL is manufactured, used or manipulated (**Regulation 12**). Prevention and removal of solid residues in connection with the use of HFL is an important means of reducing the risk of fire. **Regulation 13** requires removal of any deposit liable to give rise to a risk of fire, reasonably practicable steps having been taken to prevent the build-up of deposits. Residues containing cellulose nitrate are not to be removed using iron or steel implements, because of its comparative ease of ignition.

No smoking is allowed anywhere HFL is present and there is a risk of fire. **Regulation 14** makes the occupier responsible for policing this, including the display of appropriate signs. Provisions for the control of ignition and burning of HFLs, especially in connection with their disposal, falls under **Regulation 15**, which permits burning of HFL outside proper plant or apparatus when done as part of fire fighting training. Inspectors are empowered to take samples for testing by **Regulation 16**.

Fire fighting equipment is required by **Regulation 17** in certain dock/warehouse premises and on construction sites where HFLs are manufactured, used or manipulated. ("Factory" premises other than these are covered by the Fire Precautions Act). Finally, **Regulation 18** places duties on employees to comply with aspects of the Regulations which prohibit certain acts, to use safety equipment and report defects in it.

The Highly Flammable Liquids and Liquefied Petroleum Gases Regulations 1972

REVISION

The Highly Flammable Liquids and Liquefied Petroleum Gases Regulations provide safety requirements in the following areas:

- Storage
- Marking of store rooms, tanks, vessels etc.
- LPG – storage and marking of tanks
- Precautions against spills and leaks
- Controlling sources of ignition

- Prevention of escape of vapours, and the dispersal of dangerous concentrations
- Explosion pressure relief of fire-resisting structures
- Means of escape in case of fire
- Prevention and removal of solid residues
- Smoking
- Control of ignition and burning of HFL
- Power to take samples
- Fire fighting
- Duties of employees.

The Safety Representatives and Safety Committees Regulations 1977

INTRODUCTION

The desirability of a co-operative approach to health and safety in the workplace has been known for many years. A main recommendation of the Robens Committee in 1972 was that an internal policing system should be developed, whereby workforce representatives would play an active part in drawing hazards to the attention of workers and management, and play a positive role in explaining health and safety requirements to employees. Provision for these Regulations was made in the Health and Safety at Work etc. Act 1974, which originally contained two-subsections dealing with the subject of employee rights to consultation by the appointment of statutory representatives with whom the employer would be required to enter into dialogue.

One of the subsections allowed for the election of such representatives from amongst the workforce, the other gave the right to nominate safety representatives to Trade Unions. The provision for the general election of such representatives was repealed by the Labour government's Employment Protection Act 1975, and the right to appoint safety representatives is now restricted to recognised independent Trade Unions. (An exception is contained in the equivalent Regulations made for offshore installations in 1989, which again allow for free election). The 1977 Regulations provided the detail of the general entitlement contained in Section 2 of the Act, and came into effect on 1st October 1978. They were further amended by the Management of Health and Safety at Work Regulations 1992.

OBJECTIVES OF THE REGULATIONS

The Regulations and their accompanying Approved Code of Practice provide a set of entitlements to consultation to nominees of recognised independent Trade Unions. They are given the right to make a number of kinds of inspection, to consult with the employer, and to receive information on health and safety matters. The Regulations also provide for training and time off with pay to carry out the functions of safety representation.

HOW ARE THE OBJECTIVES MET?

The right to appoint safety representatives is restricted to Trade Unions recognised by the employer for collective bargaining or by the Arbitration and Conciliation Advisory Service (ACAS) (**Regulation 3**). The presence of only one employee belonging to such a Union is sufficient to require the employer to recognise that person (upon application by his Union) as a safety representative. The Regulation places no limit on the number of such representatives in any workplace, although the associated Approved Code of Practice and Guidance Notes observe that the criteria to consider in making this decision include total number of employees, variety of occupations, type of work activity and the degree and character of the inherent dangers. These matters should be negotiated with the appropriate Unions and the arrangements made for representation should be recorded.

Unions wishing to make appointments of safety representatives must make written notification to the employer of the names of those appointed, who must be employees of the employer except in extremely limited circumstances (**Regulation 8**) which include membership of the Musicians' Union and actors' Equity. Upon appointment in this way, safety representatives acquire statutory functions and rights, which are set out in **Regulation 4**. The employer cannot terminate an appointment; the Union concerned must notify the employer that an appointment has been terminated, or the safety representative may resign, or employment may cease at a workplace whose employees he or she represents (unless still employed at one of a number of workplaces where appointed to represent employees).

Safety representatives are not required to have qualifications, except that whoever is appointed should have been employed for the preceding two years by the employer, or have two years' experience in "similar employment". The right to time off with pay during working hours for safety representatives in order to carry out their functions and to undergo 'reasonable' training is given in **Regulation 4(2)**. What is "reasonable" in the last resort can be decided by an Industrial Tribunal, to which the representative may make complaint (see below).

The Management of Health and Safety at Work Regulations 1992 (MHSWR) have inserted a further Regulation. **Regulation 4A** requires employers to consult safety representatives in good time, in respect of those employees they represent, concerning:

- Introduction of any measure at the workplace which may substantially affect health and safety
- The employer's arrangements for appointing or nominating 'competent persons' as required by MHSWR Regulations 6(1) and 7(1)(b)
- Any health and safety information the employer is required to provide to employees
- Planning and organisation of any health and safety training the employer is required to provide

The Safety Representatives and Safety Committees Regulations 1977

- Health and safety consequences of the introduction of new technologies into the workplace.

Regulation 4A(2) requires every employer to provide such facilities and assistance as safety representatives may reasonably require to carry out their functions, which are given in Section 2(4) of the main Act and in the body of the Safety Representatives and Safety Committees Regulations.

Safety representatives have the functions of representation and consultation with the employer as provided by Section 2(4) of the Health and Safety at Work etc. Act 1974, and the following:

- Investigation of potential hazards, dangerous occurrences and causes of accidents at the workplace
- Investigation of complaints by employees represented on health, safety or welfare matters
- Making representations to the employer on matters arising from the above
- Making representations to the employer on general matters of health, safety or welfare
- Carrying out inspections of the workplace regularly, following notifiable accidents, dangerous occurrences or diseases, and documents
- Representing employees in workplace consultations with inspectors of the appropriate enforcing authority
- Receiving information from those inspectors in accordance with Section 28(8) of the main Act
- Attending safety committee meetings in the capacity of safety representative in connection with any function above.

To assist in carrying out the functions, employers must provide "such facilities and assistance as the safety representatives may reasonably require", and may be present during inspections. The facilities required include independent investigation and private discussion with employees.

INSPECTIONS

Regular routine inspections of the workplace can be carried out by entitlement every three months, having given reasonable previous notice in writing to the employer **(Regulation 5)**, or more frequently with the agreement of the employer. Where there has been a "substantial change" in the conditions of work, or new information has been published by the Health and Safety Executive relative to hazards of the workplace, further inspection may take place regardless of the time interval since the previous one. Defects noted are to be notified in writing to the employer; there is a suggested form for the purpose.

Safety representatives have a conditional right to inspect and copy certain documents by **Regulation 7**, having given the employer reasonable notice. There are restrictions on the kinds of documents which can be seen, which include commercial confidentiality, information relating to an individual unless this is consented to, information for use in legal proceedings, that which the employer cannot disclose without breaking a law, and anything where disclosure would be against national security interests.

The employer has to make additional information available so that statutory functions can be performed, and this is limited to information within the employer's knowledge. The Approved Code of Practice contains many examples of the kind of information which should be provided, and the information which need not be disclosed.

'DUTIES'

Although the Regulations give wide powers to the safety representative, they specifically impose no additional duty. Representatives are given immunity from prosecution for anything done in breach of safety law while acting as a safety representative. It has been suggested that circumstances where this immunity might apply would include agreement during consultation on, for example, a system of work proposed by the employer which later turned out to be inadequate and became the subject of prosecutions of individuals involved in the decision to use the system.

TRIBUNALS

Safety representatives may present claims to an Industrial Tribunal if it is believed that the employer has failed to allow performance of the functions laid down by the regulations, or to allow time off work with pay to which there was an entitlement (**Regulation 11**). If the Tribunal agrees, it must make a declaration, and can award compensation to the employee payable by the employer. There is no right of access to the Tribunal for the employer who feels aggrieved or who wishes to test in advance the arrangements he proposes to make.

SAFETY COMMITTEES

Despite the title of the Regulations, their only reference to safety committees is in **Regulation 9**, which requires that a safety committee must be established by the employer if at least two safety representatives request

The Safety Representatives and Safety Committees Regulations 1977

this in writing. The employer must post a notice giving the composition of the committee and the areas to be covered by it in a place where it can be read easily by employees. The safety committee must be established within three months of the request for it.

The Approved Code of Practice and the Guidance Note contain information and advice concerning the structure, role and functions of safety committees, which should be taken into account by non-statutory safety committees as well.

REVISION

The Safety Representatives and Safety Committees Regulations:

- Allow recognised Trade Unions to appoint safety representatives without limit on numbers
- Permit them to make inspections regularly, after incidents and changes in work
- Permit the inspection and copying of certain documents

- Require the employer to provide reasonable facilities for inspections
- Provide for time off with pay for safety representatives to carry out their functions
- Impose no duty on safety representatives
- Provide a detailed consultation mechanism between employers and employees
- Require a safety committee to be set up when two or more safety representatives request this.

The Safety Signs Regulations 1980

INTRODUCTION

A safety sign is one which gives a message about health and safety by means of a combination of geometric form, safety colour and symbol or text (words, letters, numbers) or both. there are four kinds of safety sign in general use:

Prohibition – (certain behaviour is prohibited)
Warning – (gives warning of a hazard)
Mandatory – (indicates a specific course of action is to be taken)
Safe condition – (gives information about safe conditions).

Supplementary signs may be used in conjunction to provide additional information. These Regulations were amongst the earliest designed to harmonise UK health and safety provisions with the rest of Europe. They are also significant in that they were among the first to identify a British Standard directly as a requirement to meet for compliance.

The Regulations partly implement Directive 77/576/EEC on the provision of safety signs at places of work. Both the British Standard and the Regulations allow exceptions, the most noteworthy of which except fire fighting, rescue equipment and emergency exits. The Regulations came into force fully on 1st January 1986.

OBJECTIVE OF THE REGULATIONS

The Regulations provide the legal means to require that safety signs comply with BS5378:Part 1:1980, where the signs are directed at people at work. The principle is that health and safety information should be presented to employees in a uniform and standardised way, keeping the use of words to a minimum. The reasons for doing so are that international trade and travel has increased, resulting in a need for international uniformity especially because of the development of international workforces which do not share a common language.

HOW IS THE OBJECTIVE MET?

Regulation 3(1) provides that safety signs must comply with BS5378:Part 1:1980, as must warning stripes in alternating colours. "Safety signs" are defined as those combining shape, colour and pictorial symbol to provide specific health and safety information or instructions. Safety signs are restricted in general to providing the instructions listed in Appendix A to the Standard, except for training use. The Regulation does not apply to road signs, marine and air traffic, or loads carried, in addition to the exception noted in the Introduction above.

Regulation 4 requires the use of normal road traffic signs to be used for regulating internal works traffic. The duty to comply with these Regulations is given to the employer or self-employed person by **Regulation 5**, but where no control can be exercised by them compliance is required from the owner of the premises. Compliance in a mine or quarry is a duty of the manager.

THE STANDARD

The British Standard was issued to coincide with the Regulations, and has been amended several times since then. The main function of the Standard is to lay out design requirements for safety signs, and to control the pictogram or symbol used in each.

Safety strips can be of yellow or fluorescent orange/red with or without a safety sign, but not in substitution for a sign in Appendix A. This contains a list and diagrams of 23 examples of safety sign. Additional signs have been designed using the principles laid down, but it should be established that all those affected are agreed on the significance of any new symbol (pictogram) before it is put into use. BS 5378:Part 3 contains specifications for additional signs to those given in Part 1.

Amendment AMD 5483 of August 1987 explained that since international agreement had not yet been reached on a symbol for emergency exits, a symbol for these has not been specifically included in the Standard. "In the meantime, however, the principles recommended in Part 1 of the standard should be adopted in relation to such a symbol."

In 1982, Part 3 of the Standard gave additional information on sign design, and added 24 more signs. Designs of symbols are to be as simple as possible, and non-essential details are to be omitted. Certain signs must show the nature of the danger in their pictograms, mandatory signs are to show only what has been mandated, and prohibition signs are to show only what has been prohibited. Supplementary text, if present, is to be clearly associated with the sign, and lettering used is to be (preferably) Helvetica Medium.

Part 3 is also concerned to illustrate with examples the principle that where more than one message is to be conveyed, separate signs must be used. Composite signs with more than one pictogram superimposed are not permitted, but multipurpose signs containing the

The Safety Signs Regulations 1980

TABLE: REQUIREMENTS FOR SAFETY SIGNS

Safety colour	Warning purpose	Examples of use	Contrasting colour	Symbol colour	Description of sign
Red	Stop Prohibition	Stop signs Identification and colour of emergency shut-down devices	White	Black	Circular red band and cross bar Red to be at least 35% of sign area
Yellow	Caution, risk of danger	Identification of hazards (fire, explosion, chemical, radiation etc.) Warning signs Identification of thresholds, dangerous passages, obstacles Risk of collision	Black	Black	Triangle with black band Yellow to be at least 50% of sign area
Blue	Mandatory action	Obligation to wear PPE Mandatory signs	White	White	Circular blue disc Blue to be at least 50% of sign area
Green	Safe condition	Identification of safety showers, first-aid points, emergency exit signs	White	White	Green square or oblong Green to be at least 50% of sign area

(The Standard sign – white arrow on a green square is not to be used alone as a safety sign.)

Supplemental text may be added if required, in the contrasting colour on the safety colour, or as black text on a white background.

appropriate symbols (and text if required) are acceptable if properly designed.

A new Safety Signs Directive (92/58/EEC) was adopted on 24th June 1992, the ninth individual Directive on worker health and safety. It will replace the existing Directive (77/576/EEC), and will require employers to use a safety sign where there is a risk which has not been adequately controlled by other means. 'Safety signs' will be extended to hand signals, marking of traffic routes and illuminated signs. The number of signs will be extended, including those to identify fire-fighting equipment, and the Directive also requires sign marking of pipes and vessels, controlling cranes and storage of dangerous substances. The Directive must be imple-

mented in the UK by 24th June 1994, with a transitional period of a further 18 months for existing signs.

The following British Standards give guidance on safety signs in addition to BS5378:

BS 5499:Part 1 – contains (statutory) requirements for signs on fire precautions and means of escape

BS 2660 – specification for exit signs internally illuminated

BS 4218 – specification for self-illuminating exit signs

BS 6304 – public information symbols

BS 4803 – signs for laser equipment

REVISION

The Safety Signs Regulations require:

- All signs giving information by their shape, colour, symbol to comply with BS5378:Part 1
- Fire fighting, rescue equipment, emergency exits excluded from this

- **Four** types of safety sign:
  Prohibition
  Mandatory
  Caution
  Safe condition.

Module D Section 28

The Ionising Radiations Regulations 1985

INTRODUCTION

These Regulations were fully implemented on 1st January 1986. They implement the majority of the provisions of EC Directives 80/836 and 84/467 Euratom, prescribing the basic safety standards for the health protection of the general public and workers against the dangers of ionising radiation. They are only concerned with ionising radiation arising from a work activity, not from the natural background, such as cosmic radiation, external radiation of terrestrial origin and internal radiation from natural radionuclides in the body. (See Section 9 of the Occupational Health and Hygiene Module for further information.)

Radiation protection is based on three general principles which are incorporated into the Regulations:

a) Every practice resulting in an exposure to ionising radiation shall be justified by the advantages it produces

b) All exposures shall be kept as low as reasonably achievable

c) The sum of doses received shall not exceed certain limits.

OBJECTIVES OF THE REGULATIONS

The primary aim of the Regulations is to introduce conditions whereby doses of ionising radiation can be maintained at an acceptable level. Details of acceptable methods of meeting the requirements of the Regulations are contained in an Approved Code of Practice (COP 16). The limiting of doses is done by setting limits on the amount of doses received in any calendar year by various categories of people liable to be exposed.

HOW ARE THE OBJECTIVES MET?

Regulations 1 – 5 deal with the interpretation of terms and some general requirements. In addition to defining the terms used in the text (**2**), such as "ionising radiation" (gamma rays, X-rays or corpuscular radiations which are capable of producing ions either directly or indirectly) and "radioactive substance", and explaining the scope of the provisions (**3**), they require employees and the self-employed to notify the Health and Safety Executive of work with ionising radiation (**5**). (There are some exceptions to this requirement.) Co-operation between employers is required concerning the exchange of information if employees of other employers are exposed (**4**).

Regulations 6 and 7 contain requirements on dose-limitation. Every employer is required to take all necessary steps to restrict so far as is reasonably practicable the extent to which employees and other persons are exposed to ionising radiation (**6**). Limits are also imposed on the doses of ionising radiations which employees and other persons may receive in any calendar year (a twelve-month period beginning on 1st January) (**7**). The limits are specified in **Schedule 1** to the Regulations.

Regulations 8 – 12 control the ways in which work can be done using ionising radiation. They specify that areas in which persons are likely to receive more than the specified doses of ionising radiation are to be designated as "controlled areas" or "supervised areas" (**8**), and entry into these is restricted to specified people in specified circumstances. Those employees who are likely to receive more than the specified doses are required to be designated as "classified persons" (**9**).

Radiation protection advisers must be appointed by employers using ionising radiations, and 'radiation protection supervisors' (**10**). These people have the duty of making local rules for the safe conduct of the work, ensuring the work is properly supervised (**11**). The advisers and supervisors may be from external organisations, working under contract to the employer. Employers must ensure that adequate information, instruction and training is given to employees and other persons potentially affected by the work (**12**).

Regulations 13 – 17 deal with dosimetry and medical surveillance. Exposures to ionising radiation received by classified and certain other specified persons are to be assessed by one or more dosimetry services approved by the Health and Safety Executive. Records of the doses received are to be made and kept for each such person.

These Regulations also require certain employees to be subject to medical surveillance, and enable the Health and Safety Executive to require employers to make approved arrangements for the protection of any individual employee.

Regulation 18 contains provisions requiring any radioactive substance used as sources of ionising radiation to be in the form of a sealed source. Any articles embodying or containing radioactive substances must be suitably designed, constructed, maintained and tested.

Also required is the accounting for (**19**), keeping (**20**) and transportation (**21**) of radioactive substances. In certain cases washing and changing facilities are to be provided. Only respiratory protective equipment (RPE) which is approved is to be used in the work, which has to be regularly examined and properly maintained. **Regulation 22** specifies washing and changing facilities in controlled areas to be provided and maintained.

Regulation 23 deals with personal protective equipment (PPE). Any PPE provided must comply with any legislation implementing provisions on design or manufacture with respect to health or safety in any relevant Community Directive applicable to head protection. Before choosing PPE, an employer must make an assessment to determine whether it satisfies Regulation 6.

The assessment must involve:

- the definition of the characteristics necessary to comply with Regulation 6
- comparison of the PPE available with the above characteristics.

The assessment must be reviewed if there is reason to suspect it is no longer valid, or if there has been significant change in the work to which it relates. Every employer must ensure that appropriate accommodation is provided for PPE when it is not being used, and employees must take all reasonable steps to return any PPE issued to the accommodation after use.

Regulation 24 requires radiation levels to be monitored in controlled and supervised areas, and requires the maintenance and testing of monitoring equipment.

Regulation 25 introduces the **concept of assessment**, placing a duty on employers undertaking work with ionising radiation to assess the potential hazards of the work. Where more than the specified quantities of radioactive substances are involved in the work, the assessment must be sent to the Health and Safety Executive (**26**). In certain circumstances employers must make contingency plans for dealing with foreseeable incidents (**27**).

Any incidents or cases involving an overexposure to employees must be notified to the Health and Safety Executive (**Regulations 28, 29 and 30**), together with those incidents where more than specified quantities of radioactive substances escape, are lost or are stolen (**31**). Investigations are necessary where employees are exposed above specified levels. For over-exposed employees, modified dose limits are provided.

Regulations 32 – 34 impose duties on manufacturers and others, and installers, to ensure that articles for use in work with ionising radiations are designed, constructed and installed so as to restrict so far as is reasonably practicable exposure to ionising radiation. Similar duties rest with employers in relation to equipment used for medical exposures.

Employers are required to investigate any defect in medical equipment which may have resulted in the overexposure of those undergoing the treatment. Interference with sources of ionising radiations is prohibited under this Part of the Regulations.

Regulations 35 – 41 deal with miscellaneous provisions, including defences available to prosecutions under the Regulations, transitional and other incidental provisions applying to offshore installations, and modifications relating to the Ministry of Defence and others.

REVISION

The Ionising Radiations Regulations 1985 contain requirements on:

- Dose limitation
- Regulation of work with ionising radiation

- Dosimetry and medical surveillance
- Control methods for radioactive substances
- Monitoring of ionising radiation
- Assessments and notifications
- Safety of articles and equipment.

The Control of Pesticides Regulations 1986

INTRODUCTION

Pesticides are chemical substances and certain micro-organisms such as bacteria, fungi or viruses which are prepared and used to destroy pests of all types, including creatures, plants and other organisms. The term "pesticide" extends to cover herbicides and fungicides, and for the purposes of these Regulations the term also encompasses a wide variety of pest control products including wood preservatives, plant growth hormones, soil sterilants, bird or animal repellents and masonry biocides as well as anti-foul boat paints.

OBJECTIVES OF THE REGULATIONS

The Regulations provide a framework for the legal control of the use of pesticides in the United Kingdom. They are designed to protect people and the environment, and enable informed official approval of pesticides as well as control of their marketing and use. The control strategy is detailed within the body of the Regulations, with a series of detailed requirements contained in attached Schedules. They were first introduced on 6th October 1986 and have been implemented progressively since that date. They were fully implemented on 1st January 1989.

HOW ARE THE OBJECTIVES MET?

The Regulations apply to all pesticides used in agriculture, forestry, food storage, horticulture, animal husbandry, wood preservation, domestic gardens, kitchens and larders. They also apply to those pesticides used in public hygiene and pest control (including parks, sportsgrounds, waterways, road and rail embankments), masonry treatment and vertebrate chemosterilant products. They extend to cover anti-fouling paints and surface coatings used on boats, on structures below water and on nets or floats/apparatus used in fish cultivation.

Exceptions within the Regulations exclude from the requirements pesticides administered directly to farm livestock (for example, as in sheep dips) nor do they apply to those used in industrial or manufacturing processes. The Regulations do apply when pesticides are used in the workplace other than in the industrial process, such as the use of herbicides on industrial land, or rodenticides in a factory. Pesticides which are not covered by the Regulations are listed in **Regulation 2(2)**, and include those used in paint, water supply systems and swimming pools.

The Regulations prohibit the advertisement, supply, storage, and use of pesticides unless they have been approved (**Regulation 4**). As a result, all manufacturers, importers and suppliers of pesticides must obtain approval for each product they develop or market for use in the UK.

The approval may be given in the form of a provisional approval or a full approval (**Regulation 5**). An experimental approval may be granted for pesticides which are being developed but these cannot be sold or advertised. The Regulations also empower Ministers to impose conditions on approval and review, revoke or suspend the approvals that have been given in appropriate circumstances.

In the event of a breach of the Regulations, powers are granted for the seizure or disposal of the pesticide (or anything treated with it), its removal from the UK (in the case of imported pesticides) or any other remedial action as may be deemed necessary. It is also possible under Regulation 8 for Ministers to make available to the public evaluations of study reports submitted on pesticides which have been granted provisional or full approval. Commercial use or unauthorised publication of information made available under this Regulation is prohibited.

More specific conditions are set out in the **four Schedules to the Regulations**. These were established over a period of five years, with all controls being fully implemented by 1st January 1989. The Schedules stipulate that:

- Only approved pesticides (including anti-fouling paints and surface coatings) may be supplied, stored or used. Only provisionally or fully approved pesticides may be sold or advertised (and then only in relation to their approved uses). Printed or broadcast advertisements must mention the active ingredients and include the phrase "Read the label before you buy: use pesticide safely". More specific warning phrases may be considered appropriate during approval and these would have to be included.

- Only pesticides specifically approved for aerial application may be applied from the air. Detailed rules are established for this specialised type of application in Schedule 4.

- General obligations must be complied with by all sellers, suppliers, stores and users, including the general public. They must take all reasonable precautions to protect the health of humans, creatures and plants and the environment. In particular, users must take precautions to avoid the pollution of water.

- Sellers, suppliers, storers and commercial users must be competent in their duties, and commercial users must have received adequate instruction and guidance in the safe, efficient and humane use of pesticides. A recognised certificate of competence is required for anyone who sells, supplies, stores or uses pesticides in agriculture, horticulture and forestry. Anyone engaged in these activities who does not hold a certificate must be under the direct supervision of someone who does. Certificates are also required for anyone using pesticides who was born after 31st December 1964, unless they are working under direct supervision.

- All users must comply with the conditions of approval relating to use. These will be clearly stated either on the label or in the published approval for the pesticide, and usually will include directions as to the protective clothing to be worn, limitations of use, maximum application rates, minimum harvest intervals, protection of bees, and keeping humans and animals out of treated areas.

- No person is allowed to use a pesticide in the course of business unless he has received adequate instruction and guidance in the safe, efficient and humane use of pesticides and is competent for the duties he will carry out. Employers have particular responsibility to ensure that every employee required to use a pesticide is provided with the necessary instruction and guidance to enable him to comply with the Regulations and achieve the standard of competence recognised by the Ministers.

- Any conditions of approval relating to the sale, supply and storage of pesticides must be complied with. Tank mixing of pesticides is controlled with lists of approval published by the Ministers.

- Pesticides approved for agricultural use (which includes agriculture, horticulture, forestry and animal husbandry) can only be used in a commercial service (applying them to other people's property or premises) by a holder of a certificate of competence or by someone under the direct supervision of a certificate holder.

The Health and Safety Information for Employees Regulations 1989

INTRODUCTION

Health and safety information is given to employees in a variety of ways. Ensuring that employees are aware of the contents of the employer's health and safety policy statement is perhaps the most obvious of these. However, earlier enactments had required that copies of printed abstracts of the major Acts and Regulations dealing with workplace health and safety to be displayed at or near the workplace in addition. Their ability to inform was doubtful, as they were simply displays of the legal requirements and not specially written for comprehension. These Regulations repealed most of the former specific requirements, and introduced a statutory set of approved written information material in the form of a poster or leaflet which must be made available to all employees regardless of the nature of their work.

OBJECTIVES OF THE REGULATIONS

The Regulations require information relating to health, safety and welfare to be furnished to employees by means of posters or leaflets in the form approved and published by the Health and Safety Executive. Details of the appropriate enforcing authority and employment medical advisory service are also made available to employees by this means. The Regulations came into force on October 18th 1989.

HOW ARE THE OBJECTIVES MET?

Regulation 4 requires employers either to provide each employee with the approved leaflet, or else to ensure that for each employee while at work the approved poster ("Health and safety law – what you should know" – no reference number) is kept displayed in a readable condition at a reasonably accessible place, and so positioned as to be easily seen and read. Any later revisions of the poster or the notice have to replace the previous ones, so out of date material is not in compliance.

Regulation 5 requires the poster to contain the name and address of the appropriate enforcing authority for the premises, and the address of the employment medical advisory service for the area in which the premises lie. Any necessary changes to either addresses or areas have to be made within six months of the change. If the employer has chosen to give the required information by leaflet, he must also give the employee a written notice containing the foregoing name and addresses. For the purpose of establishing which name and addresses are relevant for employees working away from a base, they are treated as working at the premises where the work is administered. If the work is administered from more than one base, the employer may choose any one of them as "the base".

Exemption certificates can be issued by the Health and Safety Executive under **Regulation 6** to anyone, covering any or all of the requirements, provided that they are satisfied that the health, safety or welfare of persons likely to be affected by the exemption will not be prejudiced as a result. Conditions and time limits can be attached to exemption certificates to ensure this. **Regulation 7** provides a defence of due diligence for the employer who can prove he took all reasonable precautions and exercised all due diligence to avoid the commission of an offence against the Regulations.

REVISION

These Regulations require that an information poster or leaflet is available for all employees.

The Social Security System and Benefits

INTRODUCTION

The principle that an injured or sick worker unable to earn a full wage should receive financial support has been recognised in the United Kingdom since 1837. In that year, the first case was recorded of a servant suing his employer (Priestley v. Fowler [1837] 3 M & W 1) – he lost, and in so doing established the doctrine of **common employment**. This stated that an employer could not be held liable to his servant for an injury caused by the negligence of a fellow servant with whom he was engaged in common employment. The case was supported by later decisions, and the doctrine was not abolished until 1948 by the Law Reform (Personal Injuries) Act.

In those early days, binding decisions emerged which further restricted the worker's right to claim against his employer. There was an inability to sue at common law if the injured party had also made a claim under the Workman's Compensation Act (this was not amended until 1946, by the National Insurance (Industrial Injuries) Act), and if there was any element of contributory negligence by the employee (amended by the Law Reform (Contributory Negligence) Act 1945.

Because of these reforms, financial support for those injured at work by accident or disease can be obtained by using the social security system of benefits, and by claiming damages under the tort of negligence. A **tort** is defined simply as "a civil wrong". Other Sections of this Module outline principles of the employer's and occupier's duties at common law, which depend upon the establishing of negligence on their part for the success of the claim.

The social security system is governed by specific legislation, and pays benefits to those entitled to them according to defined rules and without reference to the question of liability or fault of the parties involved.

SOCIAL SECURITY

The present social security benefit system began in 1948; funds for the system are derived from taxation, particularly of employers and employees. At first, benefits were paid for injury, disablement and to widows of those killed at work, regardless of the extent of their personal contribution to the benefit scheme.

Injury benefit was for loss of earnings where there was absence from work for more than three days following an accident in the course of employment. Circumstances which count as being "in the course of employment" are defined in Sections 50-55 of the Social Security Act 1975. Injury benefit became combined with sickness benefit in 1983 (the Social Security and Housing Benefits Act 1982). The principal Acts to consult on benefit matters generally are the Social Security Acts of 1986 and 1989.

Disablement benefit as a lump sum or pension is paid usually in respect of a prescribed occupational disease causing a loss of faculty or amenity, and which derives from employment.

The Social Security (Industrial Injury) (Prescribed Diseases) Regulations 1985 contain a list of diseases which are prescribed in relation to particular occupations or work. If a disease is not prescribed, the sufferer will not be able to claim disablement benefit but will have to sue the employer for negligence in order to obtain compensation.

Successful claims for disablement benefit must show the person is suffering from a prescribed disease and that they were in insurable employment at the time it was contracted. The claimant must also show that the disease was contracted in a prescribed occupation linked to the disease. The term "disease" in these circumstances is something of a misnomer, in that the current list of prescribed diseases contains injuries as well, including noise-induced hearing loss and tenosynovitis.

Statutory sick pay is paid by the employer for up to 28 weeks' absence, and injured employees may qualify for this.

Self-Assessment Questions and Answers

Self-Assessment Questions and Answers for Module A

SECTION 1

1. **Select a machine you are familiar with, and identify the ways in which the controls have been designed to prevent danger.**

2. **In what circumstances would plastic sheeting not be an adequate material to use in machine guarding?**

Material used in machine guarding needs to be strong enough to withstand vibration and wear caused by the normal functioning of the machine, and not be flexible enough to allow the guard to be bypassed by operators or others, or by parts, offcuts etc. which may deflect the guard from the inside. Plastic sheeting of adequate thickness and stiffness may be able to meet these requirements, but particular circumstances where it would not would depend upon the machine's characteristics and the task(s) being done.

Heat and excessive vibration, and the presence of ultraviolet light would be circumstances where plastic sheeting is not suitable. This is because ultraviolet light attacks plastic over time, heat can deform it, and vibration can cause cracking in the plastic or at the attachment points.

SECTION 2

1. **What checks should be made before a crane is used in a workplace?**

Checks should be made on the crane itself, the method of operation proposed, and the work environment. The crane must be properly installed, including a strong base having regard to the loads likely to be lifted and the capacity of the crane. The crane structure and chassis must be sound and adequately tested, and reliance may be placed upon the seller to certify this if it is a new piece of equipment. The ability of the parts to take the loads and lift safely is most important, and this can be verified first in documentation, then by a load test at installation and at intervals afterwards. The safe working load established by a competent person should be marked prominently on the structure. Indicators supplied should give the operator basic information about the load, as a minimum. Visual inspection of the lifting parts should be done frequently.

Important aspects of the crane's method of operation include the selection of and training for operators, a survey of potential hazards such as obstructions in the possible path of the crane, overhead wires, and isolation procedures during maintenance. Accessibility during regular maintenance will have to be established. Lifting equipment to be used in conjunction with the crane will need to be reviewed for capacity and suitability (slings, chains, ropes etc.).

2. **A powered lift truck overturns and the operator is injured. List the potential causes of the overturning.**

Powered trucks can overturn because of overloading, inability of the truck to deal with ground conditions under the circumstances at the time, and speed. The following factors are relevant:

Overloading by exceeding the truck's maximum capacity, travelling with the load elevated or tilted, travelling downhill with the load in front of the truck, and a shifting load.

Ground conditions include striking against obstructions or running across uneven sections of floor, turning on or crossing slopes at an angle, unsuitable floor or road for the use of the truck.

Speed includes sudden braking or changes in speed, and turning while changing speed abruptly.

SECTION 3

1. **What factors should be considered by management before the manual lifting of weights is authorised?**

Manual lifting of loads should be avoided completely where reasonably practicable. Failing this, an assessment must be made of the hazards and risks, and preventive measures considered. These steps can include alteration to the task or by mechanisation. Assessment of the hazards involves an examination of the process to note the stages at which manual handling need not be used or can be limited. Before manual handling is authorised, a safe system of work will be required which provides solutions to the remaining hazards. Such a system will include training for those carrying out manual handling and their supervisors, and also adequate health screening to ensure fit and healthy workers are selected for the task.

2. **Identify areas in your workplace where mechanised handling techniques could be used instead of manual handling.**

SECTION 4

1. **Complete the boxes on the attached sheet, identifying the common failure modes associated with each type of access equipment.**

2. A scaffold collapses as a result of overloading. What are the ways in which this could have been prevented?

Overloading of scaffolds can be prevented by limiting the loads to be placed on the scaffold, ensuring the correct type of scaffold has been specified and erected, making those working on it aware of its load limitations, and by making arrangements for its competent inspection before use and at intervals to identify any weakness which may develop. Inspection is also required to ensure the scaffold is properly supported and constructed to withstand the expected loads, and complies with any special design requirements.

Mobile towers are especially prone to overturning, often because of overloading combined with inadequate base dimensions for the working height. Stability of all types of scaffold is assisted by tying to adjacent fixed structures or by the fitting of raker support tubes to increase base dimensions.

SECTION 5

1. A vehicle maintenance fitter insists he is not exposed to danger when carrying out his work. Identify the hazards to which he is exposed.

Maintenance fitters are exposed to hazards from the vehicles on which they work, from the work environment, from work tools, equipment and procedures, and from hazardous substances they work with or become exposed to.

Vehicle hazards include working under non-propped bodies, where wheels are not chocked, brakes applied or the vehicle not adequately stabilised. Jacks should be used in conjunction with axle stands. Hot work near fuel tanks requires special procedures. Inflation of tyres at high pressure should be carried out in cages because of bursting potential, especially on split rim wheels.

Work environment hazards include noise from engines and tools, and exposure to radiation from welding equipment.

Work tools and equipment hazards include risk of eye injury from use of abrasive wheels and wire brushes, explosion of battery charging systems, hand injuries from common tools and equipment, and electrical hazards from equipment and circuits.

Hazardous substances may include dust from brake drums which may contain asbestos, fumes and gases such as carbon monoxide from exhausts and welding fumes, solvents, and degreasing chemicals.

2. The elements of a planned approach to transport safety are listed here. For your work place, make notes about any which require further attention, planning or control.

- Driver selection
- Driver training
- Driver supervision
- Control of visiting drivers
- Workplace traffic control
- Accident investigation
- Maintenance procedures.

SECTION 6

1. Find and fill in examples of each of the categories of classification for substances in your own workplace. If none are present for a category, consult a reference book and find a new example apart from those mentioned in the text.

- EXPLOSIVE / FLAMMABLE
- HARMFUL
- IRRITANT
- CORROSIVE
- TOXIC
- CARCINOGENS / MUTAGENS / TERATOGENS
- AGENTS OF ANOXIA
- NARCOTIC
- OXIDISING

2. Can you think of other, simple, ways in which substances could be classified in your workplace so as to give an indication of their potential for harm? Are there any drawbacks to your classification?

Substances could be classified in-company, using a simple coding system to indicate their assessed potential for harm:

1. Life threatening in all circumstances of exposure
2. Life threatening in some circumstances, serious injuries are probable results of exposure
3. Serious injuries probable in some circumstances, minor injuries are probable results of exposure
4. Minor injuries are the only likely outcome of exposure
5. No injuries are likely to result from exposure.

Some drawbacks to this system are:

1. It is easily applied only if data exists for each substance about its likely effects in the circumstances of use.

2. It takes no account of repeated exposures at low doses.
3. Hazardous substances generated in the workplace are likely to be missed when making classifications.
4. The system does not provide enough positive information to workers to enable them to take sensible decisions about precautions.

SECTION 7

I. Is operator training sufficient by itself to prevent chemical accidents?

Operator training is only one aspect of the safe control of chemicals at work. It is therefore necessary, but not sufficient in itself. What is required is identification of all (chemical) hazards, assessment of the risks in practice, control of these, and then operator training in the selected control techniques. Final parts of the control system are the monitoring of effectiveness, and necessary record-keeping.

2. Give an example of a hazardous substance with different degrees of risk when stored and when used.

Dynamite is such a substance. In practice, it is kept in controlled conditions, including lowered temperature, and away from the means of detonating it. In storage it is therefore a low risk, which rises as the physical conditions become less controlled when use is near, and operating procedures are relied upon for control of risk thereafter.

SECTION 8

I. What preventative measures can be taken against electrical failure?

Electrical failures and interruptions are caused by poor or damaged insulation, overheating, earth leakage currents, loose connections, inadequate circuit and component ratings, poor maintenance, inadequate systems of work and human mistakes.

These can be minimised by inspecting, testing and improving earthing standards; introducing safe systems of work including standard methods of testing for lack of power to circuits before starting work and not working on live equipment where possible; use of insulators, fuses and circuit breakers; installation of residual current devices; and use of competent workers only on electrical systems.

2. An electrical shock occurs as a result of using a drill outside. What factors might have contributed to this accident?

- Use of higher voltage than necessary for the task.
- Inadequate earthing.
- Use of unsuitable equipment – wrong drill (electric), faulty drill (not subject to maintenance?), unsuitable cable (not waterproof?), damaged cable (jointed wrongly?).
- Failure to maintain drill, cable, sockets, plugs.
- Failure to train or instruct the worker using the drill, the supervisor(s), or the management.

SECTION 9

I. How would you handle a fire involving a flammable gas cylinder?

Foam or dry powder extinguishers can be used on flammable gas fires. Cylinders containing propane or butane come into this category. Water should be used to cool a cylinder if it is leaking, because it is necessary to shut off the gas supply at the cylinder as soon as possible. If this is not done, re-ignition can occur.

2. List the design features handling fire protection in your work area.

Self-Assessment Questions and Answers for Module B

SECTION 1

1. List the possible sources of health hazards in your workplace.

2. Operatives in your workplace report general discomfort when working with a material. Discuss how you would assess the problem.

The first step is to find out as much as possible about the material. This information should be provided by the manufacturer or seller, if the material is brought into the workplace. Chemical information can be found in reference books or data bases if the material is a production byproduct. This enables the hazard to be recognised. Measurement of the hazard is then done, by finding out what standards are applicable in relation to normal exposure to the material. The information for this is likely to be available already from the first step.

The third step, evaluation of the risk in practice, is carried out by examining the method of use, handling, storage, transport and disposal of the material. Control of the risk(s) by design, engineering, elimination, substitution etc can be carried out with this knowledge. Control may also include the introduction of revised systems of work, personal protective equipment and other techniques. The final requirement of assessment is to keep in mind the need for review as conditions change, and arrangements should be made to monitor the chosen control methods at intervals to ensure that they are still effective and that the risks have not altered.

SECTION 2

1. Describe how the body can defend and repair itself when the skin is cut.

As soon as the skin is cut, blood flows and cleans the area of the cut. It rapidly clots and coagulates, which slows or prevents germs from getting into the blood or surrounding tissue through the cut. Any foreign material entering through the cut will be attacked by white blood cells, which can deal with it by chemical actions or by absorbing it. The presence of these cells is encouraged by the release of histamine in the area of the cut, which dilutes local blood vessels and increases blood flow to the area.

Repair of the area of the cut involves the removal of dead blood cells and other tissue, and the creation of scar tissue to repair physical damage.

2. Give examples of reflexes which take part in body response to the presence of foreign substances.

The coughing reflex is produced by irritating the lining of the respiratory tract – the airway from the outside into the lungs. This reflex stimulates the diaphragm to expel most of the air in the lungs in a violent movement, which forces air at speed through the respiratory tract and helps clear any particles stuck on the walls or floating in the air within the tract.

Vomiting and diarrhoea are reflexes of the gastro-intestinal tract, the pathway for food and drink through the body from entry to exit. These reflexes stimulate violent contractions of the muscular walls of the tract, which then expels substances or quantities which the body is not able to deal with by conventional means.

SECTION 3

1. "Ingestion of toxic chemicals is a rare method of contracting industrial disease". Discuss this statement, giving examples from your own experience.

2. Outline the measures which can be taken to prevent outbreaks of dermatitis.

Dermatitis is often a result of use of materials which produce an abrasive effect on the skin, or which remove natural skin secretions and allow the skin surface to dry and crack. Examination of materials in use which have the potential to cause dermatitis is a necessary first step to take.

Workers should be subject to regular simple inspection of hands etc if they are at risk of dermatitis. Good personal hygiene, practised by informed workers, is the best defence in the absence of alternative materials. Use of barrier creams and protective clothing are two other methods which are part of the personal hygiene routine, but the choice of steps to take should not be left to individuals by management.

SECTION 4

1. What are the sources of occupational exposure limits?

Occupational exposure limits are recommended or compulsory national and international standards for airborne contaminants, including most gases. In the UK, the Health and Safety Executive publishes limits annually in the Guidance Note EH40, and as necessary. In the

Self-Assessment Questions and Answers for Module B

USA, the American Conference of Governmental Industrial Hygienists (ACGIH) publishes annual lists of threshold limit values (TLVs) and the Occupational Health and Safety Administration (OSHA) publishes national standards on the recommendation of the Research Section of NIOSH, the National Institute for Occupational Safety and Health.

2. **Choose two substances found in your workplace air as contaminants, and look up their OESs or TLVs.**

SECTION 5

1. **Explain the advantages of monitoring air quality standards using stain detector tubes.**

Stain detector tubes are cheap and easy to obtain. The pumps in which they are used are also cheap, and are simple measuring devices which are unlikely to go wrong. Detector tubes give fast indication of the presence of a contaminant, and a measurement can be made easily without any special skills except in the simple method of taking a sample. When used, the tube is easy to dispose of. Only the general atmospheric level of a contaminant can be measured this way, and the contaminant must be one of those for which stain detector tubes are available.

2. **Previously, you found exposure limit values for two substances present in your workplace air. Find out what methods are, or can be, used to measure how much of each is present.**

SECTION 6

1. **Differentiate between local exhaust ventilation and dilution ventilation.**

Local exhaust ventilation traps contaminants at or close to their origins, and removes them through specially-built ventilation systems before they enter the breathing zones of workers. It uses a small hood or a ventilated enclosure or booth.

General or dilution ventilation uses fresh air to dilute a contaminated atmosphere. Although there is a general need for a regular supply of fresh air to a workplace, this should not be used as a control measure for removing contaminants unless there is only a small amount to be removed, which is evenly distributed in the workplace and is of low toxicity.

2. **Identify the various environmental control strategies employed in your workplace.**

SECTION 7

1. **Explain how noise-induced hearing loss is caused by noise at work.**

Sound is both rapid pulsations in air pressure produced by a vibrating source, and the auditory sensation produced by the organ of hearing. The mechanism of the organ of hearing is sensitive to the pulses. These are passed to it by bone conduction, and by the hearing system. This begins with the collecting action of the outer ear which captures sound, then the mechanical vibration of the ear drum produces movements of the three middle ear connecting bones. They pass the pulses into the fluid contents of the organ of hearing, the cochlea. The fluid's motion causes hair cells to rub against a membrane, which causes an electrical discharge in the hair cells. This is passed to the auditory nerve and received by the brain.

Any interruption in this process, by the body as a defence mechanism or as a result of wear or injury by excessive sound, will result in hearing defects. There are several ways in which the hearing system can fail. Noise-induced hearing loss is usually associated with the wearing out of the hair cells in the organ of hearing. This condition is not reversible.

2. **What general controls are used in your workplace to reduce noise exposure?**

SECTION 8

1. **An oxygen-deficient atmosphere has been established. Discuss the forms of RPE which should be used a) for continuous work in the area, b) for short term entry only.**

In oxygen-deficient atmospheres, RPE must provide a continuous supply of air. There are three types of equipment commonly available which will do this. The choice will depend upon the circumstances of the work to be done, including its duration, degree of difficulty and accessibility required.

Fresh air hose apparatus is relatively inexpensive, and can supply air from an external source by pump bellows or by breathing action. This would be suitable for continuous work, provided that mobility is not required, that high demand for air is not likely, and that the hose is not longer than 10m.

Compressed airline apparatus involves more equipment than the above, has the same mobility restriction, but may be useful if the outside air is contaminated or if a high air demand is required.

The final type of RPE which can be used here is self-contained breathing apparatus, which is independent of external supply or control. It has a relatively short duration of use before refilling of the cylinder(s) is required, and it is heavy, so it would be very suitable for short-duration entry where mobility is needed but less suitable for continuous work.

A more detailed analysis of the circumstances would be needed before a choice can be made of equipment for continuous work.

2. List the protective equipment in your workplace. Can you identify its limitations and suggest alternative controls?

SECTION 9

1. Explain the difference between ionising and non-ionising radiations.

Ionising radiation is a form of energy which causes the ionisation of matter which it interacts with. The energy of the radiation is sufficient to dislodge electrons from matter to which it is exposed, and in the case of the human body this can produce tissue changes. Examples are alpha, beta, gamma and X-rays.

Non-ionising radiation does not cause ionisation and its effects, although it can have other negative consequences for the body. Non-ionising radiation includes the electro-magnetic spectrum between the ultraviolet and radio waves, and artificially-generated laser beams.

2. What control measures would be appropriate to ensure safety and health of employees using or working near a microwave oven?

Microwaves are non-ionising radiations, produced artificially in ovens, where the heat of the absorbed energy cannot be dispersed so that the temperature rises proportionately with the inability to absorb the heat.

The most important control measure is to ensure, by inspection, testing and original purchase verification that the microwave oven complies with current national laws and/or standards, particularly those relating to output power restrictions, door seals and warning lights. The warning lights and door seals then require inspection and testing at least annually, depending upon use.

Training for users is another control measure which is appropriate. This should include knowledge of the function of the door seals and the need to maintain them

properly, also in the need to use a dummy load when testing (water in a suitable container) to absorb unwanted energy. Other necessary knowledge is the need to check the warning light indicating operation is functioning on each use, and not to put metallic or metal-finished objects inside the oven.

As a precaution against leaking seals, the oven should not be operated close to a permanent work station or seating area.

SECTION 10

1. Identify and list examples where ergonomic principles could be applied with benefit to your workplace. What improvements would you expect to result?

2. How can knowledge of ergonomics help in assessing manual handling risks?

Important elements of the assessment of manual handling operations are: task, load, individual characteristics and the environment in which the task will be carried out. A knowledge of ergonomic principles can help in each of these, with the result being to fit the task to the worker and not the other way around.

Understanding the limitations on human performance, designing handling equipment for safe operation, effects of extremes of cold or heat and preferred load parameters for ease of handling are examples of the practical uses of ergonomic principles.

Self-Assessment Questions and Answers for Module C

SECTION 1

1. **Explain the difference between hazard and risk, using examples from your workplace.**

2. **Write a short summary of an accident with which you are familiar, and list the immediate cause(s) and the indirect cause(s).**

SECTION 2

1. **What are the advantages of a works safety committee? Are there any disadvantages?**

A works safety committee provides a means of consulting with employees, and encourages their involvement. The committee should have a composition of roughly equal numbers of employees and management, with alternating chairmanship; it also needs a written constitution defining its scope, direction and activities, the authority it possesses (if any) and the role of its members. Debate and discussion on application of local or national standards and Regulations will be helpful in achieving their acceptance. Solutions to problems may also be generated by the committee, and problems themselves may be discovered by or brought to the attention of the committee.

Disadvantages can arise when the committee is not properly guided by expert advice, when members do not have a positive role to play, and when the committee is seen by its members and others as merely a complaints forum with no positive function other than to record problems which do not receive management attention.

2. **How is the success of your safety management programme evaluated? Are positive measures used as well as negative ones?**

SECTION 3

1. **What information will be needed before an assessment is carried out?**

Risk assessment relates the hazard under consideration to the particular circumstances. Therefore information about the circumstances (including numbers exposed) is especially important. This can be obtained by consultation with employees, studying of records and statistics in the workplace and also nationally-reported trends, and studying inspection and audits records. If appropriate, a fresh inspection including necessary measurements should be carried out to ensure the information is up to date.

The legal requirements of the situation must be available, also any guidance issued and relevant Standards.

2. **Thinking about qualitative and quantitative risk assessments, explain the difference with examples, showing where it would be appropriate for each to be used.**

Qualitative risk assessments are based upon personal judgment and general statements about the level of risk. They are appropriate in circumstances where compliance objectives do not require a more rigorous approach, such as in assessments to comply with the Management of Health and Safety at Work Regulations 1992.

Quantitative risk assessments contain probability estimates of failure, based upon observed and interpreted data. They will be required where a more precise view of risk needs to be taken, such as where public concern is raised over planning proposals, for example nuclear power stations.

SECTION 4

1. **Why are safety policies useful in the management of health and safety? Why would one consisting of only one page be of no value?**

Safety policies are useful because they a) demonstrate management intentions, support and commitment, b) set out safety goals and objectives, c) delegate responsibility and accountability for safety, c) supply organisational details, d) supply information about arrangements made for health and safety and e) they comply with any relevant statutory requirements.

The test of any safety policy will therefore be the extent to which it meets the needs of the organisation in these respects. It may be possible, in the very smallest organisations, to contain the safety policy in one page. By cross-referencing to larger documents containing detailed information, larger organisations may be able to use a single page as a basic summary of safety policy. However, if there is no cross-referencing, a single page could not contain the information and detail which is needed.

2. **Obtain a copy of your organisation's safety policy. Can you think of any improvements which might usefully be made?**

Self-Assessment Questions and Answers for Module C

SECTION 5

1. **Can you think of tasks in your workplace which require correct steps to be taken whilst carrying them out to ensure worker safety? Are they covered by written documentation? Can you justify those which are only given orally?**

2. **Write out a safe system of work for moving filing cabinets between offices.**

There are five steps involved in devising a safe system of work: task assessment, hazard identification, definition of the safe system, its implementation and monitoring the results. This task requires the use of the first three of these.

Task assessment.

Manual handling operation – task cannot be automated or mechanised, handling aids to be used. Moving (25) cabinets down one set of stairs, across a yard, up one set of stairs to new office. No mechanical handling equipment is presently available. The cabinets are all full, each drawer is full, there are four drawers to each cabinet. The maximum weight of each drawer is 13.62 kg (30 pounds), the maximum weight of each cabinet is therefore 54.48 kg (120 pounds) plus the weight of the frame – 20 kg (44 pounds). Cabinets must be unlocked prior to removal of any drawers. Doorways and stairs are wide and in good condition. There is some light traffic in the yard.

Task	Hazard	Safe method
Minimise loads before move	Attempts may be made to lift too much	All cabinets must be unlocked before work begins
Moving heavy cabinets	Manual lifting (back etc strains)	Staff selection required (fit, build, no history of injury, trained)
	Time pressure to complete may force errors in handling	Staff must not be limited by time on this task/allow adequate time
	Weight of cabinets	Do not move full or part-full units Staff to be instructed to carry each between two, use sack trolleys where possible
Remove drawers	Weight of drawers	As above
Carrying drawers	Obstructions in path	Select door props and use before lifting, **not** fire extinguishers, or assign staff to open and close them
Carrying cabinets	As above	As above
Crossing yard	Passing traffic	Limit traffic, use cones/barriers or assign staff to control traffic
Reassemble units	Put in wrong place (doors, passages may be obstructed)	Confirm proper placing with local supervisor before reassembling
	Overtipping	Confirm presence of anti-tilting mechanism on each cabinet. If not, place wooden batten under front edge of each

Other notes. All concerned must wear protective gloves and safety shoes. Assess for suitability. The task must not be proceeded with if a problem occurs which is not covered by this safe system of work. Alternative solutions to existing problems, and solutions for new problems, are not acceptable unless verified by the Safety Department.

Self-Assessment Questions and Answers for Module C

SECTION 6

1. What topics should be covered in induction safety training to provide necessary knowledge and skills for fire prevention?

Fire prevention induction training covers:

- Identification of alarms (tones, number of rings etc.)
- Action to take when alarms sound
- Basic evacuation plan
- How to raise the alarm
- Use of available fire appliances
- Fire prevention – "no smoking" rules
  housekeeping requirements
  special fire hazards of the work or workplace
  correct use of fire doors.

2. "Managers don't need to know about safety rules – they are not at risk". Discuss.

Regardless of the question of risk, managers are also employees, and have duties under common law and statute law to obey safety rules, as well as to manage their observance by employees in their charge. Equally importantly, employees will take a lead from management, so the personal actions of managers reflect the importance which management as a whole gives to safety rules. Management commitment is an essential ingredient of any safety programme – the commitment must be active and this requires the support to be demonstrated.

The risk of injury for managers is not negligible – managers appear in safety statistics, especially in the construction industry. They are as vulnerable as workers to the making of errors of personal judgment, to forgetfulness and to taking short cuts in the interests of expediency. Managers are often injured as a result of the actions of others – eye injuries in machine shops can happen to anyone present, not just the operators. Wearing of safety glasses in this situation is therefore important for everyone, as risk is no respecter of status.

SECTION 7

1. Outline the maintenance system which would lead to the rapid repair of a non-working machine guard.

Notification of defective guard is made by the operator to the supervisor, and the machine is removed from use.

(Variances from this to be authorised only by departmental manager, in writing.)

A written record to be made by the supervisor of the defect, noting defect, date and time. A copy of this goes to the maintenance department, with a priority rating to ensure prompt action is taken.

An entry to be made by the supervisor in the machine's log, or other data base. Regular occurrences of the same fault will be reported and referred for management action.

A priority system for safety rectifications is set up in the maintenance department.

There is a special budget set aside for safety maintenance work.

Supervisor has a follow-up system in place to ensure maintenance is carried out within a given time-scale.

Supervisor makes contact with the maintenance department if the time-scale is exceeded.

Ongoing regular inspections are made of machinery and guarding facilities (this may be a statutory requirement).

All concerned require training in the system, which should also be noted in the safety policy or manual.

An audit system should look at the practical operation of the system at intervals.

2. How are repairs made to the roof of your workplace? Can the maintenance system be made safer?

SECTION 8

1. Summarise the arrangements you make to select safe contractors in your workplace.

2. Develop a prequalification questionnaire suitable for use by contractors such as window cleaners, caterers and repair firms which, on completion, would help you evaluate their safety performance.

All of the steps given in the text can be shortened as appropriate, depending upon the risks of the work. Some contractors, such as window cleaners, engage in a limited range of high-risk activities, so the questionnaire for them should concentrate upon perceived problems which may arise from their work.

Self-Assessment Questions and Answers for Module C

PREQUALIFICATION QUESTIONNAIRE

1. In what type(s) of work are you skilled?

2. Who has responsibility at Board level in your organisation for health and safety?

3. Who provides you with surveillance and advice on health and safety, and what are his/her qualifications?

4. What health and safety training have you done in respect of the management/supervision you will provide, and of the operatives you will provide?

5. What is your system for investigating and reporting accidents, diseases and dangerous occurrences?

6. Has your organisation won any safety performance awards?

7. Is your organisation a member of a recognised safety organisation?

8. What are your systems for the maintenance of plant, tools and equipment?

9. Have you any written safety procedures/manuals?

10. How do you prequalify your own sub-contractors on health and safety?

11. Have you supplied us with a copy of your safety policy and specified risk assessments?

12. Please supply details of all lost-time injuries recorded to your employees during the past three years.

13. Is there any further information which would assist us in assessing your organisation's safety performance?

SECTION 9

1. **Find out the frequency rate for your workplace, and compare it with the national average for your industrial classification. Can you calculate frequency rates for individual causes of injury? Is this knowledge useful for you?**

2. **Is "carelessness" an adequate sole conclusion on an accident report?**

As a description, "carelessness" can mean almost everything – and almost nothing. Accidents are caused by human failings and by failures or the absence of control measures. Accident reports should address both of these if appropriate, not merely the human element (which is often easier to identify, especially if blame is to be laid -whose fault was it?).

"Human failings" covers all aspects of personal behaviour which result in mistakes being made, failure to take care is only one aspect. It can be argued that no-one fails to take care, in the sense that a decision (conscious or unconscious) is taken before every action, which is seen by the individual at the time to be the most appropriate response, or at least the best choice of available alternatives. Some of these alternatives should be provided or indicated by the control measures available. The decision taken may be shown afterwards to have been inappropriate or incorrect, but this does not mean that care was not taken in arriving at it – merely that it was not the right or best one. Training and experience have major influences on taking good quality decisions.

It can also be said that almost no-one deliberately sets out to have or cause an accident. Making an error of judgment is not necessarily due to carelessness; therefore further investigation is required of those accidents with a "human failing" component (all of them?), and the listing of "carelessness" as the sole conclusion of an accident report is never justified. It reveals more about the person making the report than about the accident.

SECTION 10

1. **Outline the steps you take at present to find health and safety information. How could these be improved?**

2. **Explain with examples the difference between a primary and secondary source of information. Mark the following sources as primary or secondary.**

Primary sources are original documents, collections of these separately published are secondary sources. The telephone directory and this written answer are primary sources, as are personal letters and papers given at meetings. Bibliographies and summaries of papers held on data bases are secondary sources.

Your Company annual report	Primary
Full proceedings of an annual technical conference	Primary
Your shopping list	Primary
An international standard on eye protectors	Primary
A reference library	Secondary

SECTION 11

1. **List the types of safety propaganda used in your workplace. Can you think of more effective ways of communicating the safety message?**

SECTION 12

1. **On the next two pages, you will find headings for a safety inspection checklist, with more detailed subheadings. Tick the boxes in pencil to indicate the kinds of inspections now made in your workplace for each category and the standards which apply, putting a cross where none is made. Then repeat, using a pen, indicating what the ideal inspections would be for each.**

See Module C Section 12

2. **On the third page, you will find a similar set of headings, to complete those previously given. This time, the subheadings are missing and are for you to fill in as appropriate to your workplace. Then, indicate what the ideal inspections would be.**

See Module C Section 12

Self-Assessment Questions and Answers for Module D

SECTION 2

1. Write notes on the structure and functions of Magistrates' Courts, and explain the extent of the powers of the Court in relation to health and safety legislation.

Magistrates' Courts hear criminal and civil cases; most criminal prosecutions begin and end there. Health and safety cases nearly always start in these Courts, and most are completed there as well. Magistrates are selected members of the public, not required to have legal training. Their powers of sentencing include fines of up to £5,000 and/or up to 6 months imprisonment for the more serious cases, and fines of up to £20,000 for breaches of the 'general duties' Sections of the Health and Safety at Work etc. Act 1974. If they feel their powers are insufficient, the case can be referred to the Crown Court for trial and/or sentence.

2. How do European Directives affect health and safety in the UK?

Directives are documents derived by the use of complex consultation, negotiating and voting systems within the European Community, which set out standards to be achieved within each Member State after a date stated in the Directive. Therefore, each Member State must review existing legal provisions covering the subject, and pass appropriate internal legislation to meet or exceed the standards required by the Directive.

SECTION 3

1. Outline the main differences between common law and statute law.

Common law has evolved over hundreds of years, resulting in a system of precedents or existing cases of record which are binding on future similar cases unless over-ruled by a higher court or by statute.

Statute law is the written law of the land and consists of Acts of Parliament and the Rules, Regulations and Orders made within the provisions of the Acts. Acts of Parliament dealing with health and safety matters usually set out a framework of objectives and use specific Regulations or Orders to achieve them.

The penalty for a proven breach of duties at common law is to pay compensation for the injury or damage caused, after the event. The penalty for breach of statutory duty is usually a fine or imprisonment for a criminal offence, and a case may be made for such a breach even if no accident has taken place.

2. Explain the meaning of 'reasonably practicable'.

Use of this term in legislation permits an employer, or the duty holder, to balance the cost of taking action (in terms of time and inconvenience as well as financial) against the risk being considered. If the risk is insignificant, that is that the balance between the two is "grossly disproportionate", the steps need not be taken, but lesser steps must then be evaluated in the same way.

SECTION 4

1. What are the powers of an HSE Inspector? What enforcement action can be taken by an Inspector to prevent dangerous acts taking place?

Inspectors can:

- Gain access to any place of work at any time
- Seek support from the police if entry is made difficult
- Take equipment onto premises to assist them
- Carry out inspections and examinations
- Insist that areas are left undisturbed
- Take measurements, photographs and samples
- Remove equipment for test or examination
- Take statements, records or documents
- Request facilities to be made available to assist enquiries
- Do anything else necessary to carry out enforcement duties

Inspectors can issue prohibition or improvement notices, and seize, render harmless or destroy items considered to be a source of imminent danger, and prosecute offenders.

2. Summarise the employer's duties under the Health and Safety at Work etc. Act 1974.

Generally, the employer must safeguard the health, safety and welfare at work of his employees, with specific regard to the provision of a) safe plant and systems of work, b) safe handling, storage, transport and maintenance of articles and substances used at work, c) necessary information, instruction, training and supervision, d) a safe place of work and access and egress, e) a safe working environment and adequate welfare facilities.

The employer must provide a safety policy setting out how these objectives will be achieved, in writing if he employs five or more. He must also prevent risks arising from his work activities which affect the self-employed, members of the public and employees of other employers. All the employer's duties above are subject to the test of reasonable practicability. He must also establish an adequate consultation process with his workforce, possibly through safety representatives and safety committees.

SECTION 5

1. **Review your organisation's health and safety policy and make notes on improvements needed to comply with these Regulations.**

2. **Summarise the main ways in which these Regulations expand the duties of the employer under the Health and Safety at Work etc. Act 1974.**

The Regulations have been claimed to be simple extensions and definitions of the general duties of the employer under the Health and Safety at Work etc. Act 1974. However, there are several additional duties, including the requirements to record information of various kinds.

Other additional duties are include the making of assessments, management frameworks, appointment of two types of competent person, formulating procedures for action in the event of serious and imminent danger and access to danger areas, employer co-ordination, training and capability assessment, and health surveillance.

SECTION 6

1. **Consult the Approved Code of Practice to establish the minimum number of toilets for each sex required in your workplace.**

2. **Write notes on the application of the Regulations to a forestry worker in a wood.**

   Forestry workplaces which are outdoors and away from the undertaking's main buildings are excluded from the requirements of the Regulations, except for those on sanitary conveniences, washing facilities and drinking water (Regulations 20-22), which apply 'so far as is reasonably practicable'.

SECTION 7

1. **You intend to buy a new abrasive wheel machine for use at work. What checks would you make to ensure it complies with these Regulations?**

This is a new machine, so all the Regulations will apply to it, as opposed to the requirements of the Abrasive Wheels Regulations 1970, which will not with the exception of Regulation 9 and the Schedule which deal with training. Checks required include its suitability for the task, and its conformity with current relevant EC harmonisation standards. Information on these and other aspects will be provided by the supplier by virtue of his duties under Section 6 of the Health and Safety at Work etc. Act 1974. Information about noise is included.

2. **What extra steps must be taken to ensure compliance when the machine is taken into use?**

The Regulations make provision for the continued safe use of equipment, through maintenance, authorisation of the user(s) and maintainer(s), provision of information, instruction and training to the user(s) – in this case in accordance with Regulation 9 and the Schedule to the Abrasive Wheels Regulations 1970. Lighting is to be suitable and sufficient for the work. Other Regulations, such as COSHH and Noise at Work, will also apply.

SECTION 8

1. **Identify activities in your workplace where these Regulations apply. List them.**

2. **Write notes on how these requirements affect a company supplying and installing double glazing.**

A list of all manual handling tasks must be made, from fabrication through to installation. It may be reasonable to automate or mechanise some of these, but there will be occasions when the products have to be handled manually. Written risk assessments will be required to determine the preventive measures, including any requirement for protective clothing or other equipment. Account will be taken of the more general requirements of the MHSW Regulations, especially those covering capability assessment of individuals.

The Company will have to decide how best to provide information to the handlers about the weight and centre of gravity of the loads, and also the packaging weight and content.

Self-Assessment Questions and Answers for Module D

SECTION 9

1. **List the equipment in your work area to which these Regulations apply, and identify any 'users'.**

2. **Explain the difference between a 'user' and an 'operator'.**

A user is any employee who habitually uses display screen equipment as a significant part of his or her normal work. An operator is a self-employed user.

SECTION 10

1. **Identify and list the items of personal protective equipment provided in your workplace.**

2. **Write notes on the duties of employers under the Regulations.**

The employer is required to: provide PPE where risks have not been adequately controlled by other means equally or more effective; ensure suitability through assessment, and compatibility with other PPE where used; maintain and replace lost or defective items; provide accommodation for PPE; provide instruction, information and training for users.

SECTION 13

1. **Does your own workplace require a fire certificate? Can you say why? Examine it carefully, and see how the certificate is worded and what (if any) special requirements have been added by the local fire authority.**

2. **List the contents of a fire certificate. Why are they crucial to planning for fire safety in the workplace?**

The contents of a fire certificate will include the following details: the use of the premises, the means of escape in case of fire and ensuring its maintenance, the means of fighting fire, the means of raising the alarm and particular requirements relating to explosives or highly flammable liquids kept on the premises. Additional provisions may be written in to the certificate, at the discretion of the local fire authority.

SECTION 14

1. **A new substance is to be introduced into your work place. What basic steps should be taken to ensure compliance with the COSHH Regulations?**

Obtain full information about the substance, using manufacturer's data sheets, labels, reference works, Guidance Note EH40 and other appropriate sources. Examine how the substance is to be used in the work place, including all points in the user chain from bulk entry to end disposal, as appropriate. Evaluate the risks of the proposed use, and assess what, if any, control measures are necessary. Introduce the control measures. The (written) assessment and control measures should be communicated to employees and others who may be affected by means of written information, instruction and training as necessary. Exposure should be monitored together with the effectiveness of control measures where necessary.

2. **Select an area of your work place and identify substances which are defined as hazardous to health. Attempt to find examples from the operational categories of process, raw materials, engineering, cleaning, service, byproduct.**

SECTION 15

1. **What are the conditions under which live electrical working is permissible?**

There are three basic conditions to be fulfilled. It must be unreasonable in all the circumstances of the situation under consideration for the equipment to be made dead. It must be similarly reasonable for the work to be carried out live, and adequate precautions must be taken to prevent injury.

2. **What are the qualities a person must have to carry out work with electrical apparatus?**

He or she must possess the necessary knowledge and/or experience to do the work safely, or must be under appropriate supervision having regard to the nature of the work.

SECTION 16

1. **The exposure of some of your workers has been measured and found to fall between the First and Second Action Levels. What steps should be taken to ensure compliance with the Noise at Work Regulations?**

Self-Assessment Questions and Answers for Module D

There is a general duty to reduce the risk of hearing damage generally to the lowest level reasonable practicable. The establishment of the stated level also requires the following steps to be taken:

- Make noise assessments (to be made by a competent person)
- Record the results of the noise assessments
- Provide information, instruction and training to employees about the risks and how to minimise them
- Provide hearing protectors to those people who ask for them
- Maintain and repair hearing protectors provided
- Ensure that any additional noise controls (such as baffles and enclosures) are used and maintained.

2. What are the main differences between the action required to be taken when the Second or Peak Action Levels are exceeded, and when the First Action Level is exceeded?

Exceeding the Second or Peak Action Levels requires the following steps in addition to those required when the First Action Level is reached:

- Reduction of noise exposures as far as is reasonably practicable, by means other than by provision of hearing protectors
- Mark out hearing protection zones with notices
- Provide hearing protectors to all exposed
- Ensure that the protectors are used by all persons exposed

SECTION 18

1. What considerations would have to be made to determine first-aid requirements for a construction site?

Matters to be taken into consideration in determining the adequacy of first-aid arrangements include the nature of the industry, the nature of the work and its location, the likely risks involved at the particular location, the size of the site and the numbers employed there. It could be useful to arrange for the provision of central first-aid facilities by one contractor and have these made available for the use of all.

Construction is considered to be a high-risk industry, often involving sites where access to medical treatment is difficult and may take time. Under these circumstances, a more extensive first-aid facility would be called for, staffed by appropriately-qualified personnel.

2. Explain the difference between a first-aider and an appointed person.

A first-aider is a person who has received training and obtained qualifications in first-aid which are approved by the HSE, the qualifications being current and renewable at intervals.

An 'appointed person' is a person who is capable of taking charge in an emergency and in the absence of a first-aider. They should be capable of administering **emergency** first-aid, and are responsible for the facilities provided under their charge.

SECTION 19

1. For your place of work, find out the enforcing authority, and the system followed to ensure that reportable injuries are properly notified.

2. Are the following reportable under the Regulations?

1. A resident in a nursing home trips over a contractor's cable on the floor, breaking his leg.
   Answer: Yes

2. Two cars are involved in an accident on a motorway. One of the drivers, who is driving on business during normal working hours, is killed.
   Answer: No

3. An employee is unloading materials from a lorry parked in a road outside a building site. He is hit by a passing car, admitted to hospital for observation and released two days later.
   Answer: Yes

Notes